Hope and Suffering

HOPE and SUFFERING

Children, Cancer, and the Paradox of Experimental Medicine

GRETCHEN KRUEGER

The Johns Hopkins University Press
Baltimore

Printed in the United States of America on acid-free paper
2 4 6 8 9 7 5 3 1

The Johns Hopkins University Press
2715 North Charles Street
Baltimore, Maryland 21218-4363
www.press.jhu.edu

Library of Congress Cataloging-in-Publication Data
Krueger, Gretchen Marie.
Hope and suffering : children, cancer, and the paradox
of experimental medicine / Gretchen Krueger.
p. ; cm.
Includes bibliographical references and index.
ISBN-13: 978-0-8018-8831-1 (hardcover : alk. paper)
ISBN-10: 0-8018-8831-X (hardcover : alk. paper)
1. Cancer in children—United States—History. I. Title.
[DNLM: 1. Neoplasms—drug therapy—United States. 2. Neoplasms—history—
United States. 3. Antineoplastic Agents—therapeutic use—United States.
4. Biomedical Research—history—United States. 5. Child—United States.
6. Family—United States. 7. History, 20th Century—United States.
QZ 11 AA1 K94h 2008]
RC281.C4K78 2008
362.196′994—dc22 2007040655

A catalog record for this book is available from the British Library.

Special discounts are available for bulk purchases of this book. For more information, please contact Special Sales at 410-516-6936 or specialsales@press.jhu.edu.

The Johns Hopkins University Press uses environmentally friendly book materials, including recycled text paper that is composed of at least 30 percent post-consumer waste, whenever possible. All of our book papers are acid-free, and our jackets and covers are printed on paper with recycled content.

CONTENTS

12/16/2001

Dear Journal,

I am 10 years old, and in the 5th grade. My name is Ronald Joseph Frank Voigt. I have had cancer for more than a year. This may well be two years. I'm determined to keep a dairy. Right now to January, I will be getting antibiotics. I remember how to do it, but my Mom does it. . . .

The nurse is going to come tomorrow to do my port. Sometimes, all this makes me mad, or sad, or all the others, or all of them. Riting to you makes all my tense go away. Christmas is coming up, and I can't wait. On our three-foot tree, it has a big angel on the top, lots of angels, santas and reindeers.

Well, good-bye. See you tomorrow night.*

During my postdoctoral fellowship at the Johns Hopkins University, R.J. Voigt was repeatedly admitted to the university's Children Center's Pediatric Oncology unit. Unfortunately, I never had the opportunity to meet R.J. or his family personally during their many stays, but I am grateful to have learned much about his short life and his family's heartrending tale through a Grand Rounds lecture based on "If I Die," a four-part series published in the *Baltimore Sun.* The lecture, an annual event sponsored by a family who had lost a son to cancer after extensive treatment at the research facility, was established to help teach doctors, nurses, and other medical professionals about families' experiences of childhood illness and death. As the *Sun* writer spoke, R.J.'s trips back and forth between his home on Maryland's Eastern Shore to the

*Excerpts from the journal of R.J. Voigt, *Baltimore Sun*, December 19, 2004, www.baltimoresun.com/news/health/bal-angelsjournal,0,5178698.story.

medical campus, his daily hospital routines, and, most poignantly, his alternating periods of remission and relapse—times of great hope and suffering—came alive.

R.J.'s plight was at the heart of the reporter's narrative, yet the stories and photographs vividly illustrated that he was never alone in his struggle. His mom, along with a large multidisciplinary team of technicians, therapists, social workers, teachers, nurses, pediatricians, and oncologists, shared responsibility at every phase of the boy's care. Through R.J.'s personal journals, his medical records, and extensive interviews with eighteen other families and seventy medical professionals at Hopkins, the journalist carefully untangled the complexities of modern pediatric cancer care. I am indebted to the Voigts and to all of the families who have shared their experiences so openly; to the authors and scholars who laid the groundwork for this study; to the archivists and librarians who provided access to valuable sources; and to the advisors who guided this project to completion. Without such rich sources and dedicated support, this book (and the dissertation that preceded it) would have been impossible.

Many mentors have shared their wisdom and their time. My two undergraduate advisors at Truman State University, Philip Wilson and Janna McLean, engaged me in original research and opened my eyes to the intellectual excitement that enlivens academic communities. They prepared me well for the rigor of graduate school. At Yale University, John Warner and Naomi Rogers pushed the bounds of my dissertation ever outward, and Daniel Kevles was an exacting teacher as he supervised my work in the classroom and carefully reviewed the scientific aspects of each chapter. I owe an especially large debt to my advisor and friend, Susan E. Lederer, who shared her expansive knowledge about the history of medical ethics, the use of children as research subjects, and popular health, thus helping me build a firm foundation for this project. She was also unfailingly generous with her time and insisted on high-quality writing, clear thinking, and clean prose, habits that have improved this book immeasurably and are lasting gifts that will serve me throughout my career. Others provided crucial feedback and support at this stage. Barron Lerner, David Cantor, Keith Wailoo, Janet Golden, Leslie Reagan, and countless others provided key sources, critical feedback, or a new contact at key moments during the writing process. My editor, Jacqueline Wehmueller, saw the promise of this project at an early stage and provided direction. Her involvement and encouragement propelled me from first chapter to dissertation to book manuscript.

I found another intellectual home and refuge for writing at the Institute for the History of Medicine at the Johns Hopkins University. Through the support of this community and the American Society of Clinical Oncology (ASCO), I was given three years to focus on my book project and to deepen my understanding of twentieth-century cancer research and treatment. Many colleagues, including Randall Packard, Mary Fissell, Harry Marks, Jesse Ballenger, and Deborah Whippen, helped me refine my ideas and translate them for various audiences and publications. Two vibrant communities also added new dimensions to my project at key junctures: the Center for Children and Childhood Studies at Rutgers University–Camden and an international group of historians and sociologists focused on cancer organized largely by John Pickstone at the Center for the History of Science, Technology and Medicine at the University of Manchester.

Funding from Yale University and several departmental grants facilitated my research, travel, and writing. I would like to extend my sincere thanks to the following institutions and individuals for providing assistance and valuable sources: the American Cancer Society in Atlanta, Georgia and its Media Division in New York City; Charles Balch and Lisa Persinger at the American Society of Clinical Oncology; the Columbia Oral History Collection; Bill Schaller and the Dana-Farber Cancer Institute; James Holland at Mt. Sinai Hospital; George Canellos at Dana-Farber Cancer Institute; the Lymphoma and Leukemia Society of America; the National Archives at College Park, Maryland; Special Collections of the Regenstein Library at the University of Chicago; Rockefeller Archive Center, Pocantico Hills, New York; and the Schlesinger Library at the Radcliffe Institute, Cambridge, Massachusetts. My colleagues in the Family, Business, and Medical History Center at Wells Fargo Bank enabled me to complete the final phase of editing.

I would also like to thank the *Bulletin of the History of Medicine* for allowing me to reprint sections of an article ("*Death Be Not Proud:* Children, Families, and Cancer in Postwar America" 78/4 [Winter 2004]: 836–863) that appear in chapter three. This article, awarded honorable mention by the Shryock Committee of the American Association for the History of Medicine, benefited from the judges' excellent suggestions. Chapters two and four reprint portions of "'For Jimmy and the Boys and Girls of America': Publicizing Childhood Cancers in Twentieth-Century America" 81/1 (Spring 2007): 70–93, from the same journal.

Drafting chapters and making revisions are activities best done in isolation, but writers would surely be lost without the close companionship of friends and the love of family. My parents, Rick and Jean Krueger, told me that anything was possible and helped me live my dreams. My husband, Jim Schuck, believed in me at times when I had lost faith in myself. He has cheered and sustained me from the first word to the last, despite three years living on opposite coasts. I am grateful to all who opened their homes and lent a needed hand to keep us connected across the miles.

Hope and Suffering

Introduction

In 1966, a *Life* magazine article by Will Bradbury titled "Two Views—the Lab, the Victim" illustrated the stark contrast between the promise of scientific research and the outlook for children diagnosed with acute lymphocytic leukemia (ALL), the most common pediatric cancer.[1] The story consisted of two separate, yet intertwining parts. A series of black-and-white photographs captured the story of Mike Parker, a ten-year-old boy who received urgent treatment at M. D. Anderson Cancer Center in Houston, Texas. Alongside the bleak pictures of Mike's therapy and his parents' anxious wait, vivid color photographs displayed the "all-out research assault" launched against the disease. The parallel layout pictured a life-and-death race. Although chemotherapy temporarily halted the progression of Mike's disease, readers were left wondering whether the boy would survive until the next scientific breakthrough.

The Parkers' story began in a fourth-grade classroom located sixty miles north of Abilene, Texas, where Mike had suddenly changed from being an eager student to a listless pupil. Soon thereafter, he began begging to stay home from school. After watching this unusual behavior for several days, Mike's mother, Ella, called to make a doctor's appointment for her son. The osteopathic physician diagnosed a common cold and prescribed vitamin and penicillin injections but did not suspect what was to come. A few weeks later, Ella had to rush her son to the hospital in the middle of the night with an uncontrollable nosebleed. A blood count the following morning revealed that Mike had ALL. He remained at the local hospital for six days before he was referred to M. D. Anderson for further treatment.[2]

The record of the boy's admission to the specialized cancer hospital reveals the rapid advance of Mike's disease. He had already lost sixteen pounds (a quarter of his total weight), leukemic cells had infiltrated and inflamed the bone coverings in his joints, his liver and spleen were enlarged, and tiny hemorrhages called petechiae partially obstructed his eyesight and dotted his legs. After noting this list of findings, the doctor tried to reassure the Parkers: "He's about average for a leukemic child. Not any worse."[3] Photographs captured the busy schedule of activities that began to govern his days: regular blood transfusions, chemotherapy treatments, bone marrow tests, and a strict mouth care regimen to help guard against bleeding gums and prevent dangerous infections. Mike's mother moved into her son's room to supervise his care while her husband returned home to work and look after their other two children. A bout of pneumonia caused a brief setback, but Mike successfully achieved his first remission after six weeks of continuous treatment. Despite his progress, a caption framing a photo of Mike returning home on a bus with his parents warned that the boy's journey would not end here: "Doctors at Anderson know from sad experience that he will be back."[4]

Despite the grim trajectory of Mike's story, Bradbury described the scientific segments optimistically, suggesting that ongoing efforts to identify and test new chemotherapeutic agents and new virus research programs would soon reap benefits for young patients and families like the Parkers. A striking photo taken at the Sloan-Kettering Institute for Cancer Research in New York City showcased eight chemicals that had shown activity against ALL. One agent, mercaptopurine (6-MP), had been used in Mike's case. In the background, cabinet-lined walls housed samples of the 28,000 chemicals screened for anticancer activity at the institution. Other images provided snapshots of the bench and animal experiments carried out as part of the Special Virus Leukemia Program, a "superplan" designed by the National Cancer Institute (NCI) to expedite leukemia-related virus research. As Mike received treatment at M. D. Anderson, investigators at NCI, Sloan-Kettering, City of Hope Hospital in Duarte, California, Roswell Park in Buffalo, New York, and dozens of other hospitals and research centers hoped to construct a viral hypothesis of leukemia and develop a vaccine against the dreaded disease.[5]

Mike Parker was only one of many young patients with cancer in the twentieth century caught in the uncertain space between hope and suffering. In 1930s, cancer had been redefined from a dread disease that struck only adults and the elderly to a rare but major killer of infants, children, and youth. By comparing the case records of young patients treated at the Memorial Hospital for Cancer and Allied Diseases in New York (now better known as Memorial Sloan Kettering), to extant medical literature and New York City vital statistics data, Harold Dargeon, a newly hired Sloan-Kettering pediatrician, determined that particular types of cancer disproportionately affected children while others predominated in later life. The small set of solid tumors

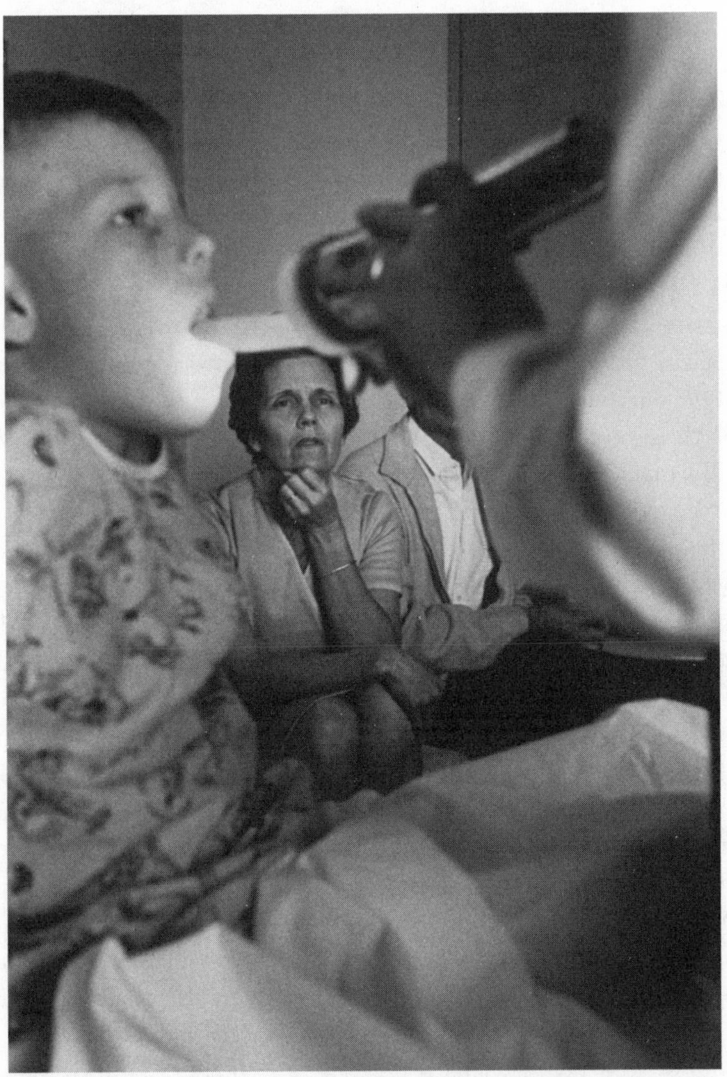

Mike Parker in the hospital. The *Life* photographer not only captured the details of the boy's busy hospital routine but also documented his parents' role in providing care. They looked on as his mouth was checked routinely for signs of infection. To protect his gums against bleeding and other complications, nurses recommended using a foaming toothpaste instead of a brush. Monitoring Mike's condition was a vital, constant part of his treatment. *Time,* November 18, 1966.

and cancers of the blood and related tissues most frequently observed in children from birth to fourteen years old were reclassified as "childhood cancers."[6] This research also led to another significant finding: mortality from cancer ranked second only to accidental death in the young. As mortality from common infectious diseases of children declined, cancer became a growing concern. The private institution established a host of child-centered programs and services during the late 1930s and 1940s in response to these startling observations. Families from the metropolitan area, across the country, and around the world began traveling to Memorial and, in later decades, other specialized cancer centers, for access to experimental therapies and enrollment in clinical trials. This fervent search for hope in the midst of suffering shaped gains in the knowledge about cancer, changes in practice and policy, and the interactions between parents and the medical professionals that began to govern the management of cancer in the young.

Attending to the complex negotiations between children, parents, medical professionals, and others involved in the care of pediatric patients and the treatment of childhood cancers, *Hope and Suffering* is a history of childhood cancers in twentieth-century America. But this story's roots reach decades earlier. Beginning in the 1850s, philanthropists founded children's hospitals in major American cities to provide a separate institutional home for the physical healing and moral education of young patients.[7] Separate children's wards were also established at general hospitals. By establishing hospital positions within these independent departments, gaining teaching appointments, and opening specialized private practices, physicians shaped a discrete specialty of pediatrics and became legitimate members of the medical community by the end of World War I.[8] Pediatricians had solidified their authority by promoting well-child visits and precise routines of proper feeding, cleaning, and consumerism termed "scientific motherhood" that established the necessity for pediatricians' expert advice in times of both sickness and health.[9] This new pattern of medical supervision then expanded when experts in child guidance and specialists in adolescent medicine promoted the importance of identifying and studying the mental and physical issues of growth and development.[10] Parental caretakers—usually mothers—played only an ancillary role in their child's care.

By the 1930s, significant attention was devoted to maintaining and restoring the health of children—inside *and* outside the bounds of a doctor's visit or the walls of a clinic. Public health reformers during the Progressive Era supported preventive measures against communicable diseases by ensuring ample supplies of pure food, milk, and water, all of which contributed to declines in life-threatening infectious diseases such as measles, whooping cough, diphtheria, and scarlet fever.[11] As rates of infant and child mortality fell, fears shifted to other causes of death. It was at this time that Dargeon joined the staff at Memorial and uncovered cancer's profound impact on children's health. Soon thereafter, childhood cancer appeared in the hospital's publications, the pages of women's magazines, and in the wider public view.

Although it had been determined that different types of cancer struck children and adults, all cases inspired both fear and hope. Cancer alarmed laypersons because of its associations with pain, disfigurement, and inevitable death. Before the advent of chemotherapy, radiation and surgery were the only conventional therapies available to treat cancer, and, as the Parker's story dramatically illustrated, the development of chemical agents after World War II did not guarantee long-term survival. In *The Dread Disease*, James Patterson described the early activities of the American Society for the Control of Cancer, the predecessor of the American Cancer Society (ACS), as it began its fundraising and educational campaigns to promote awareness and early detection through the "Seven Warning Signals of Cancer."[12] *Dread Disease* included little discussion relative to the cancer society's attention toward young persons with cancer, but it joined several accounts that explored the complex interrelationships among the popular images and understanding of cancer, political activities related to directing and funding national cancer research programs, and the development and delivery of therapeutic modalities.[13] Scholars have continued to pursue these themes, but through projects that are narrower in scope and transnational in perspective.

A growing body of scholarship has begun to evaluate the utilization of varied treatment modalities, prevention strategies, causal theories, and professional structures specific to different cancers and countries. Most important, this preliminary has revealed that a single, global model of cancer management and control did not and does not exist.[14] It will require much additional research to understand each individual system as well as the entire network. Recent sociological inquiries into the organization of cooperative cancer groups in the United States such as the Acute Leukemia Group B (a child-centered program), however, contributed important groundwork to this story.[15] In addition, complementary work on childhood cancers in the United Kingdom outlines another national structure and highlights possible points of contestation and cooperation between the United States and other countries.[16] This book focuses on children (and families) in America and their roles as part of a unique culture of clinical experimentation built around pediatric cancers in the mid- to late twentieth century in which the roles of physician, patient, and hospital became closely intertwined with those of the investigator, research subject, and laboratory.

Why focus on stories told by and about children with cancer? Historians of childhood have long recognized the ways by which children wield power as young subjects of national debates and political action.[17] As Russell Viner and Janet Golden have argued, children have also played a vital part in shaping medical history. By listening for children's voices in the historical record and searching carefully for the faint imprints of their actions we can understand the channels by which children have negotiated or altered the practice of medicine and the policies governing care.[18] By coauthoring illness narratives, creating comic books, composing diary entries and letters, and even participating in conferences, children critiqued hospital procedures, revealed their understanding of their diagnosis or prognosis, and detailed their complex negotiations with their caregivers.

Cancer institutions and organizations have also strategically used children to appeal to potential donors. Like young polio sufferers, children with cancer served as "poster children" to personify and personalize a menacing disease. Such materials played upon the twin realities of hope and suffering. As death rates fell overall, children's deaths from cancer became all the more heart-wrenching and disturbing. Young representatives poignantly illustrated the unnecessary loss of innocent victims to cancer; thereby undermining the child-centered model of the family and depriving the nation of future citizens. In contrast, poster children and children who made guest appearances on popular radio shows and in annual telethons also dramatized the hope and promise that a cure for all cancers was near. Through modern advertising and fundraising techniques, cancer—in adults and children—entered popular consciousness by the mid- to late twentieth century. Such materials add children's voices and faces to the dominant narrative—one that has given primacy to innovations in the medical management of childhood cancers.[19]

Childhood cancers must be understood as a disease that affects not only the child but also the entire family. By linking illness narratives and major "medical milestones" achieved in the laboratory, hospital, and outpatient clinic, *Hope and Suffering* reconstructs families' changing experience of disease from the 1930s through the late 1970s. Exploration of parents' involvement in their child's illness and their negotiation of the modern medical marketplace lends complexity to a history that is often told as a story of biomedical triumph by cancer organizations heralding statistics about improving survival rates, telethon hosts speaking with bald-headed children, physicians reporting the results of complex, multimodal treatment protocols, or pioneers in specialties such as pediatric oncology and hematology recounting the history of clinical research related to acute leukemia or other common childhood cancers. Top-down histories of pediatric cancers are valuable contributions to the field, but they miss half the story. Such accounts mask the lived experiences of the young patient and caregivers when confronted with a rare, but devastating disease. Correcting this omission is not only a matter of justice; it also attends to powerful forces that have remained relatively invisible in the historical record but have had an increasing role in shaping cancer treatment, policy, and funding in the United States.

Like breast cancer activists and celebrities, pediatric cancer patients and their families were not simply passive recipients of medical knowledge and care. Instead, they have been powerful arbiters of medical principles and practices.[20] Scholars who have studied childhood diseases, maternal and child welfare programs, and the medicalization of child health have argued that the image of distressed parents seeking treatment for a sick or dying child has social, emotional, and political power. *Death Be Not Proud,* a memoir of a teenage boy's illness and death from a brain tumor published in 1949, illustrated one couple's desperate search for experimental therapies at a time when multimodal treatment including surgery, radiation, and chemotherapy was first available for cancer. Thousands of letters sent from readers revealed that John Gunther, the author of the best-selling book, was not alone in this relentless pursuit; families from across the country had shared the Gunthers' experiences and identified with their pain. By establishing new organizations such as the Leukemia and Lymphoma Society of America (originally the DeVilliers Foundation), parents sponsored research programs for this unique group of cancers and, notably, memorialized their child. Parents also provided care at the bedside. The discovery of several effective chemotherapeutic agents, the rapid establishment of large-scale drug identification and screening programs, the design of combination drug protocols, the organization of a cooperative clinical trial structure, and the development of supportive therapies to control dangerous complications of cancer and its treatment led to longer survival times, especially for ALL. As many types of childhood cancer—and conditions such as juvenile diabetes, hemophilia, and cystic fibrosis—were transformed from acute, fatal illnesses to diseases that responded favorably to long-term medical management, care shifted from the hospital to the outpatient clinic and home.[21] The duties of parents expanded as they adopted primary responsibility for dispensing oral chemotherapy drugs, shuttling their child to frequent appointments, and closely monitoring the side effects of treatment.

From court proceedings to the bedside to Capitol Hill, integrating parents and other caregivers into the history of childhood cancer expands the boundaries of care and recognizes the participation of many more actors in the story of a disease. This approach uncovers the uncertainty not often expressed in top-down histories—that is, those written from the perspectives of institutions or experts—by showing that the hardships or suffering caused by an intractable disease or rigorous experimental therapies often tarnished the heady claims frequently repeated by journalists and investigators about cancer breakthroughs.

Drawing upon a wide range of published and unpublished sources, accounts of therapeutic innovation can be recast as nuanced histories of cooperation, skepticism, and resistance among patients, parents, and practitioners. Alongside excerpts from medical texts and conference proceedings, illness narratives offer detailed records of family's daily activities and their private concerns. Letters illustrate the diversity of patient and parent experience as they also illuminate common themes such as the availability of curative therapy, the merit of alternative treatments, decisions to pursue or halt therapy, or possible causes of cancer. Court records demonstrate intersections between cancer, child health, and the law. Newspaper and magazine articles, as well as fundraising and educational materials, document the stories of cancer sufferers; they also provided insight into prevailing messages about cancer, including those common to children. Campaigns often displayed childhood cancers to promote a standard slogan first promoted by the ACS that early detection and prompt treatment by orthodox physicians led to a cure, without addressing the peculiarities of this set of malignancies. These rich sources illustrate the paradox of hope and suffering that characterized pediatric cancers: childhood cancer research "breakthroughs" in clinical trial methods and drug development initiated early claims that a cure was near, but these high hopes were frustrated when sweeping cancer cures were not achieved quickly.

This book is organized chronologically in order to pair the dominant narrative of the history of childhood cancer—the transformation of acute leukemia and other common childhood cancers from inevitable killers to curable cancers—with the personal experiences of young patients and their families. Each chapter opens with a family's story, before expanding outward to consider several key themes that cross every decade: etiology, early detection, treatment, short- and long-term side effects, death and dying, and cure.

The first chapter begins in the 1930s, the decade in which physicians at Memorial Hospital for Cancer and Allied Diseases first recognized and classified cancers in children as a separate set of neoplasms and a leading cause of child mortality. Following this discovery, Memorial appointed a pediatrician, constructed additional treatment facilities, and launched a public relations campaign that featured the hospital's youngest patients. Although the hospital's child-centered initiatives garnered local attention, a series of controversial cancer cases heralded by front-page headlines made the American public more aware of certain childhood cancers and the lived experience of the dread disease. I examine the creation of "glioma babies"—young children diagnosed with retinoblastoma—and the role of the media in transforming private medical decisions into a public spectacle. In each case, when family members decided to withhold conventional cancer treatments, they were pressured to cede control of their child's health to the state, physicians, ethics boards, or other authorities. With the proliferation of experts around the care of children and their particular medical problems, childhood cancers and those affected by the disease became the targets of increased scrutiny.

The second chapter is framed by the illness experience of "Jimmy," a boy who traveled from his home in Maine to Boston to receive cancer treatment and became the first spokesperson for a new child-centered charity, named the "Jimmy Fund" in his honor. The images, voices, and stories of Jimmy and others personified and raised the public awareness of cancer in the young at a time when biomedical research efforts were receiving increased institutional interest and federal support. This chapter details the expansion of childhood cancer efforts at Memorial Hospital and other major research centers. By publicizing their youngest patients' tragic stories, work on childhood cancer was promoted as a critical area of investigation and an area worthy of research dollars.

Chapter 3 probes John and Frances Gunther's story about their son and his prolonged illness more deeply. Published in 1949, the best-selling memoir *Death Be Not Proud* offered a detailed record of Johnny Gunther's illness and the Gunthers' yearlong search to find treatments for his incurable brain tumor.[22] The volume provided a narrow glimpse into one family's life with cancer, while thousands of letters sent to Johnny's parents after the publication of the book and related magazine excerpts provided wide-ranging sources that exemplified the diversity of patient and parent experience in the late 1940s and early 1950s. The correspondence also revealed the silence surrounding childhood cancers and the manner in which the illness narrative bound together a diffuse community experiencing similar parental loss and grief. Parents sought this refuge at a time when discussion of child death (especially from cancer) was hushed and a cure was still unknown.

A single, lengthy letter written to John and Frances Gunther by Angela Burns, the mother of a daughter with ALL, serves as the basis of the fourth chapter. Burns enumerated the challenges posed by the side effects of toxic treatment, the rigors of making daily trips from their home to Boston, and her simultaneous praise and skepticism of chemotherapy.[23] By considering Burns's story alongside the development of a series of promising, yet only temporarily effective, highly toxic chemotherapeutic agents, the intertwined nature of medical science, the medical management of illness, and the family's role in caretaking is rendered in sharp relief.

Peter De Vries's novel *The Blood of the Lamb* provides the foundation for a chapter on patterns of remission, relapse, and child death caused by the introduction of new chemotherapeutic agents and rigorous combination regimens in the 1960s and 1970s.[24] The semifictional account of young Carol Wanderhope's illness and death from ALL in the early 1960s was based on the author's personal tragedy of losing his own ten-year-old daughter from the disease. Truth-telling debates, the repeated cycles of illness and wellness produced by newly developed chemotherapeutic agents, and the relocation of care from the hospital to the outpatient clinic and home are recorded and criticized in the novel.

In the 1960s and 1970s, investigators had begun to tentatively suggest that children with several types of childhood cancer—including ALL—could be labeled as "cured." Despite these advances, it remained a time of uncertainty. This final chapter examines a small cluster of cancer narratives that served as important vehicles for parents to share their varied experiences as they coped with illness and incurability, short- and long-term survival, and death and dying and sought to publicly appraise their child's medical care at a time when patient activism and parent advocacy on behalf of ill and disabled children was growing in strength.

At the end of this period, it was recognized that prolonged survival and cures had come at a price. Follow-up studies found that children often suffered lasting mental and physical disabilities from the toxic therapeutic regimens employed to kill cancer cells completely. Researchers now asked what it meant to be a "truly cured child," and parents now faced another set of concerns regarding the long-term, healthy survival of their children. The epilogue briefly considers the concerns of physicians and parents as they face the continued challenges posed by childhood cancers, their treatment, and the anticipation of a cure.

By looking through the eyes of the children and families described on these pages we can gain greater insight into the management of adult cases. How are the two sets of cancers intertwined? During the past century, the popular interest created around children with cancer has educated Americans about new views of cancer and, at times, has spurred challenges to ideologies promoted by "experts" in the field. The aggressive, fast-growing nature of pediatric cancers has also advanced our scientific understanding of the basic biological mechanisms of the disease, suggested alternate models of cell growth and regulation, revealed unexpected etiological theories, and exposed the high price of immunosuppression exacted by chemotherapy. In addition, acute leukemia served as the basis for the first cooperative cancer clinical trials—a system now central to the evaluation of all treatment protocols and a vast transnational research network. This model has not only governed cancer research since the 1950s but has also led to the establishment of new medical specialties and professional structures in the hospital and clinic, including comprehensive care and hospice. It is clear that each facet of modern cancer care has been influenced by some of the littlest cancer patients—patients like Mike Parker. By looking both to the past and the present, we can note the agency of children and their caregivers in informing policy and practice and should use these insights to reshape our ideas not only about the best care of sick children, but the impact of modern biomedicine on our own health and wellness.

CHAPTER ONE

"Glioma Babies," Families, and Cancer in Children in the 1930s

In April 1933, members of the Vasko family of Hastings-on-Hudson, New York, barricaded themselves in the back rooms of their second-story apartment. There John Vasko, his wife, Anna, and their daughters remained well concealed from dozens of newspaper reporters and curiosity seekers milling about on the lawn below. Four months earlier, Helen, one of the Vasko's two-year-old twin daughters, had been diagnosed with a malignant eye tumor—a tumor called a retinoblastoma, referred to in the popular press as a "glioma." Specialists repeatedly warned the couple that the cancer growing on the toddler's retina would spread from one eye to the other and then migrate to the brain if the affected eye were not removed promptly, but the Vaskos refused to allow surgeons to proceed with the operation. When informal negotiations with the family stalled, physicians and members of community social service agencies worked together to challenge the Vaskos' controversial position in court. As the case progressed to the appellate level, newspapers published bold headlines, front-page stories, and intimate photographs of the "glioma baby" and her family, thereby transforming the Vaskos' private decision into a public spectacle.[1] As attention to their plight escalated, the family retreated.

The Vasko case was part of a small series of well-publicized "glioma baby" cases that began making cancer in children more visible in the United States in the 1930s. Human-interest stories about the families employed sensational methods to attract the attention of local subscribers as well as more distant readers. Readers commented on the contested cases on editorial pages and expressed their divergent views through opinion polls. Reporters writing about families' struggles not only brought childhood cancers to the public eye, they caused a lively debate among physicians and laypersons over the proper medical care for children with cancer and other physical ailments.

The Human Laboratory

Helen Vasko's cancer diagnosis came at a pivotal point in the treatment and definition of this complex set of diseases. Long considered intractable and, perhaps by extension, unmentionable, physicians paid increased attention to cancer after the development of antiseptic and aseptic surgical techniques permitted the first successes in the field. According to Patrice Pinell, "Excision techniques for malignant tumors were first codified in the late nineteenth century, and cancer became a field in which a new generation of surgeons could start making a name for themselves."[2] Breast, lung, and other organ-specific cancers became the sites of surgical experimentation and innovation. The recommendation to remove the young girl's affected eye was surely grounded in the prevailing theory that timely, extensive eradication of the tumor and surrounding tissue often yielded the best outcome.

Yet, before the 1930s, few medical texts or journal articles about cancer discussed retinoblastoma or cancer in children, more generally. A surgical fellow at the Mayo Clinic described the predominant attitude: "From the time of the earliest medical writings to those of the present day, cancer has been defined and discussed as a disease of middle or late life. Malignant disease is likely to be excluded from the realm of probability when a patient gives his age as 25 or less."[3] There were, however, a few exceptions. In the late nineteenth century, several French physicians attempted to identify and quantify cancers in infants, and, although children did not receive prolonged attention in his 1908 text *The Natural History of Cancer*, Roger Williams observed, "One noteworthy feature about the tumors of infancy and early life is, that the localities whence they are prone to originate, are very different from those whence malignant tumours commonly arise at later periods of life."[4]

In the 1920s and 1930s, the few published articles on benign and malignant tumors often described single cases or small cohorts of young patients who had been diagnosed with tumors usually detected in adults, but there was a growing recognition among practitioners that there was a separate class of cancers that affected only infants and children.[5] For example, several physicians observed that a tumor of the central nervous system specifically affected young children. Although the investigators debated its proper name, they agreed on its primary site, its rapid course, and that it grew during embryonic development and then commonly first appeared during infancy.[6] Neuroblastomas (a tumor of specific nerve cells), kidney tumors (later named Wilm's tumor), and retinoblastomas were some of the first cancers that physicians suspected of affecting infants and children specifically.

It was during this decade that Memorial Hospital for Cancer and Allied Diseases shifted its attention to cancers in the young. Memorial had been established in 1884 as the New York Cancer Hospital, one of the first institutions devoted entirely to the research and treatment of cancer. Child patients, however, had not received separate care. From 1917 to January 1, 1930, 311 cases of malignant tumors in children younger than fifteen were treated at Memorial, but children were admitted to and cared for in the same wards as adult patients.[7] Beginning in the 1930s, research targeting cancers in the private institution's youngest patients led to the expansion of existing facilities and the development of new spaces and services to identify, document, and treat cancers in children.

Memorial's efforts began on a small scale but increased in scope over the course of a decade, establishing the hospital as a leader in this area of research and treatment. The 1930 article "The Age and Sex Distribution and Incidence of Neoplastic Diseases at the Memorial Hospital, New York City" provides a window into staff members' early work. The broad statistical study included brief case reports to record the appearance of certain tumors at unusual ages, particularly in the "extremes of life."[8] The article characterized Wilm's tumor as a congenital neoplasm and noted "the average of the subjects with this neoplasm was less than three years, the youngest average age among a hundred different histological varieties of tumors."[9] It also made note of the striking number of gliomas of the eye diagnosed in children younger than five. In their conclusion, the authors divided the normal human life into five age epochs (infancy, childhood, presexual period, maturity, senescence) and attempted to isolate the malignant tumors linked with each division. Again, they returned to the same two tumors, saying that they were "endowed with great propensity and capacity for growth" and that aberrations during embryonic development were responsible for their occurrence.

The survey results led to several structural changes at Memorial. In 1933, George Pack and Hayes Martin, cancer surgeons at the hospital, organized separate clinics in Memorial's mixed tumor and head and neck divisions to accommodate pediatric patients. The next year, a pediatric ward with four beds was opened, and in 1935, Harold Dargeon was appointed consulting pediatrician—the first at the institution. To emphasize "the concept of a child as a person suffering from a grave illness" rather than a disease-centered approach, Memorial's medical board chose a pediatrician rather than a cancer specialist to fill this post.[10] Although Dargeon and the establishment of the Department of Cancer in Children had the support of the board, others affiliated with the hospital viewed work with the youngest cancer patients as unrealistic and hopeless. Dargeon observed that "children almost invariably had suffered from cancer for incredibly long periods; not only months, but not infrequently years before they reached definitive treatment." He predicted optimistically that earlier, accurate cancer detection would contribute to higher cure rates when paired with appropriate radiation and surgical interventions.[11]

In 1937, the pediatric section of the New York Academy of Medicine sponsored a landmark symposium to present the findings of a comprehensive study that compared the extant medical literature on cancer in children with Memorial's case records. In *Cancer in Childhood,* the published conference proceedings, Dargeon and James Ewing, the administrator of Memorial, presented compelling evidence that a discrete set of cancers disproportionately struck children.

Ewing called for a major revision of earlier findings, concluding that researchers had relied on improper age divisions, incorrect tumor classifications, and small sample sizes when studying cancer in children. The previous scholarship commonly grouped infants and children together with young people beyond the first two decades of life and, thus, had discovered few noticeable differences between the neoplasms affecting children and adults. Ewing advised physicians making similar investigations to limit their scope to the first ten to fifteen years of life, the time of infancy and childhood, when, he hypothesized, "Special factors of heredity, nutrition, and growth may be expected to express themselves." He characterized children's cancers as those which "progress rapidly and metastasize widely" and added, "The recurrence is prompt and the mortality very high."[12] Ewing also noted that authors writing in the late nineteenth and early twentieth century had tended to group all forms of cancer—carcinomas, sarcomas, benign or malignant—into one vague category. He counseled his colleagues to be more sensitive to the differences among tumors as they redirected their attention toward the clinical course of tumors in the young. "It is clear," Ewing argued, "that the conditions of origin and the clinical course of these diseases are so peculiar that they may not be properly compared with any adult tumors, and that this entire subject deserves to be treated as a special department in the descriptive history of neoplastic diseases."[13] The leader's conclusions definitively supported the creation of a pediatric cancer research program at the institution.

Memorial's metropolitan location and its status as a leading cancer hospital helped facilitate large-scale studies comparing children's cancer cases. With the addition of a pediatrician to the staff, childhood cancers became a major focus of Memorial's prevention, research, treatment, training, and record-keeping initiatives. It was at this time of limited, yet growing awareness about pediatric cancers in the cancer research community when Helen Vasko was first diagnosed with a glioma.

Setting the Stage

Elizabeth Tomashevsky, a YWCA worker who served as an interpreter and mediator between the Vaskos—who had emigrated from Czechoslovakia—and authorities, recounted the story of Helen's diagnosis to the press. She explained that Helen's mother had first observed an unusual spot on her daughter's eye and brought her to the Hastings Village nurse for an exam in January 1933. The nurse quickly dismissed the young mother's concerns, advising her to regularly rub Helen's eye with boric acid. It is unclear what she hoped to achieve with this simple remedy, but it produced an unsatisfactory result. After two weeks of treatment, the spot had grown and the nurse referred mother and daughter to Grasslands County Hospital. There, a physician correctly diagnosed the rare tumor. The Vaskos then sought a second opinion from an eye, ear, nose, and throat specialist, who confirmed the diagnosis and warned that the tumor would surely grow inward toward the girl's brain. At this point, Tomashevsky met with the Vaskos to reiterate the physicians' grave prognoses and to reassure them a glass eye could be inserted for cosmetic purposes. Despite this assurance, the couple refused to allow the operation.

The Vaskos were skeptical, even dismissive, of the physicians' claims that their child was suffering from a life-threatening illness. Contrary to the grim diagnosis, John Vasko observed that his daughter continued to laugh and play, signs that she was not in any pain. "The doctors know nothing," he chided. "They are all crazy."[14] He chose to rely on the counsel of Basil Beretz, their parish priest at the Greek Catholic Church, as a trustworthy source of guidance. Following a private meeting between the couple and the priest, John Vasko relented, saying, "The law knows best. If law says take child's eye [*sic*], I say all right." Anna Vasko, however, insisted, "God gave me the child. If God wills it, He will take her away."[15] She set up a small religious shrine, prayed daily for a miracle to heal Helen, and purchased eye drops from a dispenser of herbal medicines to treat her daughter at home. The mother's practices were looked upon with suspicion because they were used at a time when the familiar routines and modalities of home-based care were being replaced by hospital-centered treatment and healers shifted from family members to medical specialists.[16] The couple may have also been under increased scrutiny because of the ethnic differences and economic disparities that marked them as distinct from the authorities charged with their child's health and safety. Like Mary Mallon (known more commonly as Typhoid Mary), an Irish-born cook working in New York City who was imprisoned after repeatedly exposing others to infection, members of the Vasko family found themselves under increased surveillance and threat of punishment if they did not conform to expert opinion.[17]

Anna Vasko resisted the surgical procedure, in part, because of the permanent blindness or disfigurement that could result. Although the 1930s have been described as a decade of transition from the isolation and separation of the blind to greater protection by the state and integration into mainstream American society, these achievements did little to improve the lives of blind children.[18] Education for blind children remained scattered among privately funded institutions, state residential schools, and integrated public school classrooms. Even when taught, blind children were often grouped haphazardly with others with a wide range of mental and physical disabilities. By 1940, a few blind youth had begun attending colleges and universities, but, as a historian of the National Federation of the Blind noted, "For the vast majority of those who graduated from schools for the blind, the prospects of a normal life and livelihood were almost as dismal as they had been a century before."[19] Stigmatization of children and adults affected by blindness limited their opportunities even when students earned degrees. The "glioma baby" cases illustrated that for some parents in the 1930s it was difficult to decide what was best for their child—death or permanent blindness in a society not equipped to properly educate or gainfully employ them. In New York state, however, the advent of the children's court system in 1922 opened the way for authorities to intervene in and direct parents' difficult health decisions involving children and youth to age seventeen.[20]

Entering the Court

Upon learning of the Vaskos' decision to refuse therapy, staff members at the county hospital reported the case to the Society for the Prevention of Cruelty to Children for further evaluation.[21] The society, like the police, members of men's and women's clubs, school principals, and public nurses, identified potentially mistreated children and, when necessary, intervened in their homes with the legal aid of the children's court. At this point, the society alerted Judge Smyth of the Westchester County Children's Court to the Vasko's controversial position and appealed to him to consider whether Helen Vasko was a victim of neglect or whether her parents were simply protecting her from further harm.

The children's court hearing included surprising medical testimony. Specialists consulted in the case had repeatedly and vehemently insisted on the immediate removal of Helen's eye, but they now testified that the surgery held only a fifty-fifty chance of saving the young girl's life. They stated that by the time observable signs or symptoms of the tumor appeared, the cancer had usually advanced and destroyed vision in the affected eye. By delaying the surgery still further, the Vaskos had, perhaps, allowed the cancer to grow or spread more widely. Although they knew that the chance of preserving her sight or life was increasingly slight, physicians continued to mandate the operation.

At the time this testimony was given, the published medical literature on tumors of the eye was slim but growing rapidly as several new investigators entered the field. In the nineteenth century, European ophthalmologists had broadly defined gliomas as all tumors of the retina, but in 1926 a committee appointed by the American Ophthalmological Society suggested dividing gliomas into specific categories to reflect more accurately the specific tissue from which each tumor arose. The most common type of malignant eye tumor—the one Helen suffered from—was reclassified as a retinoblastoma. Although cancers were not yet classified specifically as "pediatric" or "adult" in the late 1920s, physicians observed that retinoblastomas usually affected children under the age of five and that there was a higher incidence of the cancer in twins and siblings, lending evidence to a hereditary mode of transmission.[22]

In the late 1920s, Algernon Reese, an ophthalmologist who had completed extensive training in the pathology of the eye in Boston, New York, and Vienna, joined the staff of Memorial Hospital and began investigating potential therapies for retinoblastoma, which he characterized as "frightful."[23] He and others divided the progression of the tumor into three discrete stages: an early period, when a white or yellow reflection resembling a "cat's eye" appeared; an inflammatory stage, when the eye became irritated from increasing pressure; and a third stage, when it was thought that the tumor spread through the optic nerve to the brain. Physicians hypothesized that death resulted from this local migration.[24] Based on this model, the timely removal or enucleation of the affected eye or eyes was the only option Reese endorsed.[25] The medical experts who testified in court repeated the conclusions of this literature—if any chance for survival remained, immediate and aggressive surgery was best. The development and dissemination of expertise in this new medical subfield opened families' personal decisions to scrutiny by providing grounds for contestation.

A brief submitted by Francis Fay, the Vaskos' lawyer, used the 50 percent statistic relied upon by the specialists as a counterargument, writing, "There is no certainty an operation will arrest the malignant growth." He expressed doubt whether a "person of ordinary prudence would resort to such a dangerous operation when the result is so speculative."[26] According to Fay, the specialists' ambiguity left ample room for parental skepticism and choice.

After considering the arguments in the Vasko case, Judge Smyth favored the physicians' expert testimony when making his decision. He deemed the operation "necessary" and ordered that it be performed immediately at the local hospital. Smyth also asked the Westchester County Bar Association and the county children's association to appoint a guardian for Helen. The Vaskos quickly filed an appeal with the Appellate Division of the New York Supreme Court to keep their daughter at home and temporarily stall the procedure. By now, both the case and the Vasko family were subject to intense public scrutiny.

Under Public Scrutiny

Hearings of the children's court usually remained private, thus the Vaskos' story did not attract widespread publicity until it reached the appellate level. When it did, however, the *New York Times* and *Los Angeles Times* joined local newspapers in summarizing recent events and chronicling the daily developments in the Vasko story. Front-page headlines announced "Doomed Baby Is Barricaded in Home" and printed photographs of the crowd that had gathered around the three-family house. Frightened by the growing assembly, Anna Vasko blocked the doors and windows of their apartment and threatened to douse onlookers with boiling water. When Helen's twin suffered from a dangerous 105 degree fever, they barred the physician from making a house call for fear that he would take her sister. John Vasko stayed home from his job in order to protect his family. He complained,

> Men come and—boom—shoot big lights in my face. One of them steals the picture of my girl. They make my child Anna sick, too. Why don't people let us alone? Let them mind their own business. They're our children. Lawyer says that maybe the best thing to do is to have an operation. If the courts say so, I may say all right. But my wife don't know yet [*sic*].[27]

As Bert Hansen, Nancy Tomes, and Barron Lerner have shown in their work on the history of medicine, newspapers transformed stories of suffering and disease into compelling human-interest stories to entertain readers, encourage public involvement, and achieve commercial gain beginning in the late nineteenth century and continuing (with increased intensity) through the present day.[28] To their dismay, the Vaskos also found themselves at the center of such a project. In addition to tracking the family's activities, newspapers summarized the content of the legal proceedings and printed readers' opinions of the case. They reported that two pivotal questions faced the court: "Has the state a right to demand an operation on Helen, which probably will save her life but destroy the sight of one eye?" and "Has a parent the right to decide a question meaning possible life or death for an infant?" Readers aligned themselves with the two stances defended in court: "Humanitarians" argued that a life must be saved at any cost and that parents had no rights over the existence of a child, while "strict legalists" argued that the state had no right to encroach on the wishes of the family.[29]

This polarity was mirrored in hundreds of letters sent to the Vaskos' attorney. Although some writers were critical of the couple's "ignorance of medical science" and "misguided opposition to curative therapy," others denounced outsiders' involvement. One former Yonkers resident told of a similar personal experience and accused the experts of humiliating families: "If some of these so-called community workers would mind their own business they would be doing the community a great charity."[30] The Citizens' Medical Reference Bureau of New York City passed a resolution against such interference, calling it proof of "medical autocracy."[31] Not only entertaining human-interest stories, "glioma baby" cases had entered the court of public opinion. At an impasse between their personal convictions, expert opinions, and public outcry, the Vasko family tried to evade the media spotlight and, perhaps, the jurisdiction of the state by moving to a secret location one early morning.[32] Only the milkman saw them flee their home.

Final Deliberations

On Tuesday, April 18, the Appellate Division court in Brooklyn unanimously upheld the state's right to order the operation. The court judged Helen Vasko to be a "neglected child" who required protection by the state. One justice wrote,

> This appeal presents, primarily, the right of the State, in a proper case, to assume the discharge of duties of parents or guardian in matters involving the life, health, and physical welfare of their children or wards when it appears that the parents or guardian, through ignorance, fanaticism, or for arbitrary reasons, have become derelict in their duty and failed to perform it.[33]

The court predicted a "tragic prospect" for the child without the operation and, thus, chose medical intervention over nontreatment. A supporting memorandum cited the nature of medical knowledge as grounds for the decision. "With the world-wide recognition of the scientific character of the practice of medicine," it noted, "we come as near to a stage of certainty in result as it is humanly possible."[34] The decision helped shift power over children's health and welfare away from parents to experts.

Reaction to the opinion was swift. A piece written by the editor of a local newspaper publicly supported the involvement of the court in such cases, characterizing the legal intervention as "helpful" since it relieved the couple of the burden of making a decision. The author encouraged courts to be "zealous in their protection of insane persons and children because they cannot protect themselves," although he or she recognized that this stance could be considered a "highhanded assumption of authority."[35] Despite this risk, the Vasko decision gave physicians, social services, and the state greater license to arbitrate questions that had previously been considered to be under parental control.

Such efforts, however, only strengthened the Vaskos' resolve. Following the decision, the family again went into hiding to evade the officials charged with moving Helen to the county hospital for observation. After a weeklong search, authorities found the family and persuaded them to permit another consultation by ophthalmologists in New York City. The family was accompanied by a series of police escorts from Hastings to Yonkers to New York. This service was reportedly provided to guard them from curious onlookers, but it also ensured that they completed the trip. When the specialists' report confirmed the earlier diagnosis, Anna Vasko relented slightly, saying she was "ready to consider" the surgery. Yet, the couple remained undecided whether they would consent to the operation or urge the Court of Appeals (the highest tribunal in the state) to review the case.

Conflicting stories in the press suggested that the final interaction between the family and the state may have begun with negotiation but ended in force. The couple, Judge Smyth, the child's appointed guardian, and other members of the judge's staff met in the judge's chambers to discuss the appellate court decision and the content of the latest medical report. During the extended conference, the judge focused primarily on easing Anna Vasko's concerns that her daughter would encounter lifelong hardships as a partially blind citizen. One account reported that the judge's compassion, sympathy, and even tears persuaded the young mother to change her mind. After giving her consent, she bundled up Helen's clothing and then carried Helen to the elevator to be transported to New York City for surgery.[36] By contrast, two deputy sheriffs who observed the scene described a tumultuous conclusion to the deliberations. They reported that Anna Vasko had pleaded for permission to take her child home to Czechoslovakia. Then, when interpreters in the judge's chambers said that the mother had given her consent, Vasko would still not release the child. Finally, at the judge's order, the deputies forcibly took the baby from her mother to be transported to Columbia-Presbyterian Medical Center for surgery.[37] After her daughter was gone, Anna Vasko took to her bed and reportedly threatened to drown herself in the river if the infant died during the operation.[38]

The following afternoon, the baby's left eye and a portion of the optic nerve were removed in a thirteen-minute procedure. An official bulletin issued a few days later described Helen's rapid recovery, reporting that she was able to sit up in bed and play with her toys. Yet, it cautioned, "While the operation offers . . . the only possible chance for conserving life, this chance is not a full one."[39] A pathological examination of the removed tissue found that the disease was "fairly advanced" but had not extended into the optic nerve, physical evidence that Helen's prognosis was still favorable despite the long delay. The young girl was soon released from the hospital; she returned home to complete her recovery.

It is unclear whether Helen Vasko was cured, but *In re Vasko* left a lasting mark on deliberations over child health by becoming one of the earliest cases in which a parent's decision not to allow surgical attention was legally overturned and termed "neglect."[40] The case had an immediate impact. Within a month of the decision, a Brooklyn court considered whether the parents of fifteen-year-old Benjamin Rogalski could be compelled to allow an operation for a serious hernial condition, which could turn fatal if not corrected. Newspapers reported, "The ruling in the Vasko case gives the court the power to order the surgery if the boy's condition necessitates it."[41] The judge adjourned the case to review *In re Vasko* thoroughly before pronouncing his final decision. The cancer case also helped established a legal foundation for modern cases and law reviews exploring health issues related to minors in the twentieth century—from child abuse cases in the 1960s, to nontreatment decisions for infants in the 1970s and 1980s,[42] to debates regarding sex assignment on infants with ambiguous genitalia,[43] to recent childhood cancer cases in which parents choose to replace orthodox regimens with alternative therapies—cases that now consider young patients' voices alongside parents' responsibilities and physicians' clinical judgment.

The Vasko's story provided an intimate glimpse into one family's struggle with a little known, life-threatening disease at a time of institutional growth and the proliferation of expert knowledge around childhood cancers. It also serves as a valuable, early point of comparison to a series of other "glioma baby" cases that made local and national newspaper headlines during the 1930s. As treatment options changed and perceptions of physical disabilities were disputed, parents continued to find their decisions mediated by a growing body of experts. By contrasting one family's experience with another, it is clear that the identity of an individual family may have played a significant part in determining the "proper" care for a "frightening" disease.

Meet the Colans

On Saturday, May 7, 1938, approximately five years after the Vasko decision, the headline on the front page of the *Chicago Daily Tribune* announced a "Battle to Save Life of Baby." Only a few days earlier, Morris Hershman, a physician and the maternal grandfather of five-week-old Helaine Judith Colan, had observed the signs of a tumor in Helaine's eyes, confirmed the diagnosis with several colleagues and broke the grave news to his daughter and son-in-law, Herman and Estelle Colan. When the couple brought their baby to a nearby hospital for further evaluation, they unintentionally stirred up a public debate about the best treatment for the infant.

The outlook for young retinoblastoma sufferers remained grim in the late 1930s. The *Tribune*'s vivid descriptions of the rare pediatric cancer informed readers of the disease's mysterious cause, its "insidious progress," and its frequent recurrence after surgery:

> Glioma of the retina, the malady that threatens to take the life of little Helaine Judith Colan, is one of the dread afflictions against which medical science knows no defense other than surgery, which may fail to save the patient's life. It advances stealthily and there is no way to detect its approach. It makes itself apparent only after it is too late to stay its progress.[44]

Anatomical drawings illustrated the path by which the cancer moved inward from the eye and vivid quotes from local medical authorities emphasized the dangers of the tumor's rapid growth while promoting surgery and aggressive, experimental radiation regimens as the best course of treatment. Newspapers provided another outlet for the new medical knowledge and authority to stretch beyond the walls of specialized research hospitals to potential patients.

Physicians advised Helaine's parents of their limited options—nontreatment or the immediate removal of both eyes. In an interview, Hershman admitted that he was struggling to reconcile his professional judgment with his loving devotion to his new granddaughter. "As a physician," he stated, "I believe nature should be allowed to take its course. As a father, and as this child's grandfather, however, I am inclined to the other side—that of trying to save the baby." In the midst of this indecision, Helaine's father Herman, a dentist, soon announced the couple's choice to "shun the operation and let nature take its course."[45] This official statement, however, was not the final word.

News coverage of the controversial story did not refer explicitly to the Vaskos, but it did revisit local cases that had involved the health and long-term survival of young children. In the 1910s, H. J. Haiselden, chief of staff at the German-American Hospital in Chicago, had allowed infants with physical deformities or evidence of moral or mental defects to die. The Bollinger case had attracted widespread attention when Haiselden decided against an operation that would have saved the life of a crippled, paralytic baby.[46] Despite the outcry that followed his decision, he continued his highly criticized work until his death four years later. The Colans' life-and-death decision prompted a similar ethical debate among Chicago's leaders and citizens as well as hundreds of letter writers from across the country who followed and commented upon the case.

Public opinion polls revealed divergent views. A Gallup Poll designed to measure views on the Colan case from a religious perspective found a majority of respondents agreeing that life should be saved whenever possible.[47]

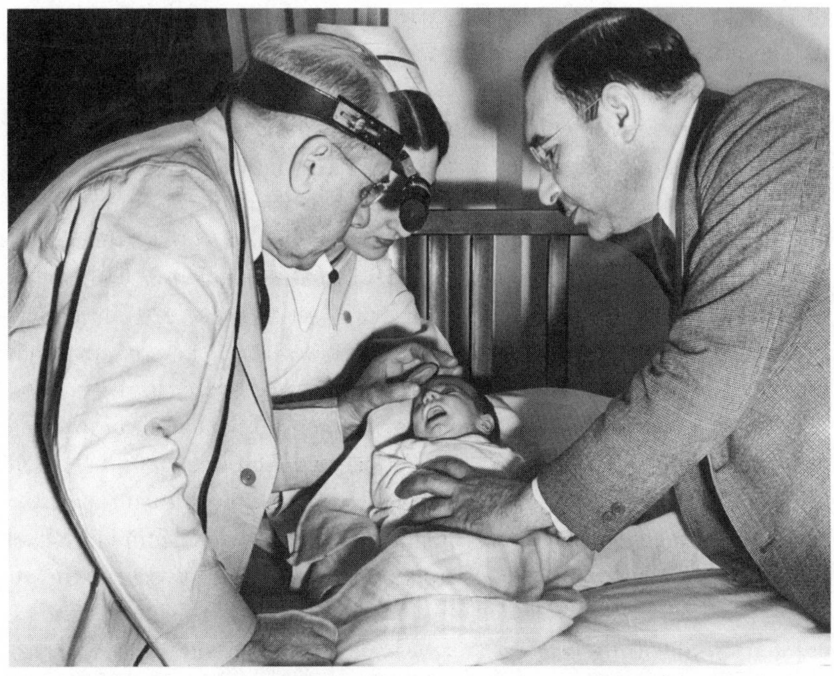

Dr. Robert H. Good (left), Nurse Lucille Rapp, and Dr. Morris Hershman, grandfather of the infant, examining Helaine Judith Colan at Garfield Park Community Hospital.

Quotes published from a range of civic leaders in medicine, business, and urban society, however, demonstrated more variability. For example, some physicians defended surgical intervention by saying that it ensured the best chance for survival, while others admitted that the surgery had an uncertain result and should not be pursued. Frank E. Adair, head of the medical board of Memorial Hospital in New York, said that most parents making a similar decision chose to let the young patient die and admitted that he would do the same for own children. Although such statements clearly demonstrated that doctors had not yet adopted a unified position, food industrialist Oscar Mayer and Mrs. Charles H. G. Kimbell, president of the Chicago Junior League, urged parents to rely on their physician's opinion when in doubt. It is clear that these prominent laypersons put their trust in medical authority, despite experts' lingering uncertainty.

As in the earlier case, commentators' perception of the blind helped shape their view of the alternatives. While several interviewees felt that the disability would be an unnecessary burden on the child, his or her parents, and society, most agreed that blindness was preferable to death. Residents of an industrial home for the blind told the Colans that schools for the blind would train their daughter for a useful life but cautioned that achieving self-sufficiency remained a difficult challenge. On a more positive note, Helen Keller telegraphed the Colans personally to urge them to give their child the chance to live and assure them that Helaine could overcome her disability. Although Anna Vasko had feared that her daughter's sight would be impaired, it was clear that Helaine would lose her sight completely with the removal of both eyes.

The *Tribune* published stories and photographs of the Colans and others coping with "glioma," illustrating the number of families facing the same disease and hard decisions. In the Chicago area alone, four other cases were diagnosed within a few days of the Colans. One, in particular, captured readers' attention. The parents of Richard Edmark also refused to consent to the removal of their twenty-two-month-old son's eyes, declaring simply, "We are doing it for his own good."[48] The newspaper also profiled several local survivors. Triumphant tales were told of young women who had partially or completely lost their sight from retinoblastoma but had survived and successfully gone on to work or marry and raise children.

Assailed by these differing views, the Colans organized a secret late-night conference with their parents, siblings, and a lawyer to make a final decision. The proceedings, described as "a mixture of science, parental love, and emotion," lasted until the early hours of the morning, yet the participants were not able to reach a consensus.[49] The family concluded that the "problem was one too difficult for those so intimately and personally concerned to decide" and opted to leave the final decision to a jury of medical specialists.

The Colan case topped the local news for three days straight as a jury of physicians from local hospitals, two "brain specialists," a pathologist, an x-ray specialist, three rabbis, and the attorney serving as a spokesperson for the family met under the direction of Irving S. Cutter, dean of the Northwestern Medical School and regular health columnist for the *Tribune.* On the recommendation of Robert Good, a physician at the community hospital, they agreed on an option that had not yet been considered: the sightless, left eye would be removed and then studied in the laboratory to determine whether the other eye should also be excised or whether a series of x-ray or radium treatments could replace the second radical surgery. As soon as the family approved the decision, Helaine was wheeled into the operating room.

This last-minute option garnered consensus from the expert panel and appeased the Colan's concerns, but whether it was a viable alternative or supplement to surgery was questionable. The discovery of x-rays, radioactivity, and radium in the late nineteenth century supplied new tools against cancer. While the responsiveness of skin cancers to x-rays first gave credibility to "radiotherapy," serious concerns remained. Could doses be calibrated that were both safe and effective? How would they change as more powerful roentgen and radium therapy devices were developed or as physicians targeted tumors located more deeply in the body? In the United States, unlike Europe and Canada, surgery—not radiation—continued to be used as the primary mode of cancer treatment after World War I.[50] The organization of hospitals and physicians, the role of the governmental funding, and the priorities of voluntary cancer associations all influenced therapeutic practice in different national contexts. Without the creation of a central agency to purchase and distribute radium supplies or to standardize the use of x-ray or radiation therapy, these modalities remained secondary in America. Despite this variation, by the 1930s, preliminary evidence had demonstrated that the radiation therapy had a measurable affect on cancers of the eye, thus Good and his colleagues accepted it as a more conservative option that would prod the Colans toward action. Helen served as a well-publicized test case for the modality.

The young baby and her family remained in the spotlight after the operation. An official bulletin reported that the baby "whose fight for life is being watched by all the nation" was in stable condition.[51] In a separate statement, the Colans expressed their appreciation for the jury and their role in the process: "My wife and I are satisfied that the doctors' committee has done everything to minimize our worry, and we are eternally grateful to the doctors for their decision."[52] Even so, they selectively ignored other forms of expert guidance as their daughter's condition improved. Based on their rabbi's counsel, the couple had agreed to postpone visiting Helaine until she had made definite progress and to institutionalize her if the surgery caused brain damage. Despite this pledge, Estelle rushed to visit her daughter immediately after the operation had ended. She also amended some of her earlier statements, claiming that she now wanted to save her baby's life at any cost, even if it meant complete blindness. She insisted, "All the time all I wanted was to save her life. That was my first thought, and I didn't change."[53]

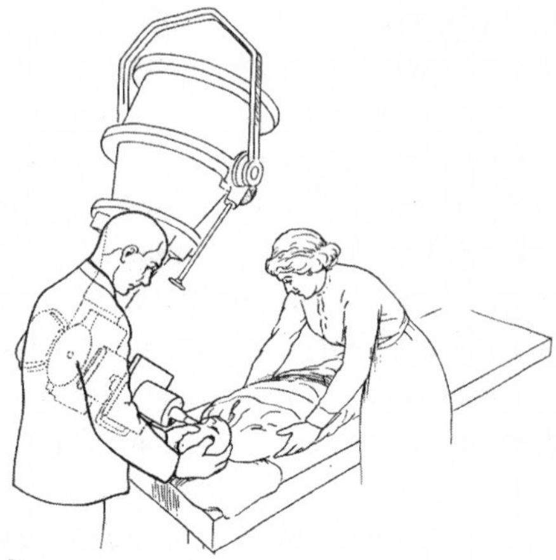

Administering radiation to children. Observation and treatment of children with cancer was dependent, in large part, on whether the child's behavior could be managed by practitioners or parents. During radiation therapy, the child's body was immobilized with a sheet. One parent held the child's body, while the other steadied the head. The authors of the article explained that the risk of x-ray exposure should be assumed by the family of the child, not a staff member who would be affected continuously if called into service for every case. Source: Hayes Martin and Algernon B. Reese, "Treatment of Bilateral Retinoblastoma (Retinal Glioma) Surgically and by Irradiation," *Archives of Ophthalmology* 33 (June 1945): 435. Reprinted with permission of the *Archives of Ophthalmology*/AMA.

At a follow-up conference, the medical jury reviewed the pathological report before concluding that it was best to try to shrink the remaining tumor and preserve Helaine's remaining sight through supervoltage x-ray treatments. The prescribed regimen included daily eight-minute treatments over a three-week period, a schedule that was later extended. In August, Estelle Colan and her mother brought four-month-old Helaine to New York City for a follow-up appointment with ophthalmologist and retinoblastoma specialist Algernon Reese. Reese observed that the tumor showed a substantial regression and declared that no further surgery was necessary. He advised, though, that additional radium treatments be administered in Chicago.

Tragically, on August 15, 1940, more than two years after her diagnosis, Helaine died from retinoblastoma and its complications. Although the young girl had appeared outwardly healthy for months, she had abruptly and completely lost her sight in the early summer of that year. Newspapers did not report whether the Colans had refused or pursued further medical intervention at that time.

Beyond Disease

The two "glioma baby" cases illustrate intersections between disability and disease, the perceived rights and responsibilities of parenthood, and the growing authority of physicians and other experts over medical decision making during a decade of marked transition from the provision of medical care in the home to technologically driven hospital care. The increased power commanded by and granted to physicians altered parents' role in their child's health and medical care. For example, pediatricians insisted on the benefits of routine well-child care, public health workers prescribed methods for the prevention of infectious diseases, and in these cases specialists dictated the proper treatment for young cancer sufferers. Whether this intervention was justified at a time of therapeutic uncertainty left ample room for discussion and disagreement about cancer. Little was still known about the cause, course, or effective therapies for retinoblastoma, yet one reporter wrote, "Eyesight, if the doctors enable her to keep it, will be in the nature of a gift from medical science to the child, rather than a birthright."[54] It is clear from this quotation and the outcome of the glioma baby cases that physicians had been granted elevated status among some Americans.

Whether a specific cancer case was resolved by contestation, negotiation, or cooperation depended, in part, on the individual family. As several historians have suggested, the "productive partnership" between parents, physicians, philanthropic social agencies, and the government that contributed to a marked reduction of infectious diseases and infant mortality in the early twentieth century should not be romanticized. In part, success hinged on experts' ability to build relationships with parents or others involved in providing care for children, a task that had to accommodate significant ethnic and class differences, issues related to geography and access to care, and a host of other factors. The Vaskos' and Colans' experiences demonstrated that intervention—even legal intervention—into parents' life-and-death dilemmas related to cancer took many different forms and produced varied results depending on a family's particular circumstances.

Although both the Vaskos and Colans combined medical science, parental devotion, and their religious beliefs when considering their child's condition and future, newspaper reporters described each family quite differently. In

New York, writers set ethnicity and class in the foreground of their articles, frequently referring to the Vasko's Eastern European roots. The Vaskos were characterized as "simple folk" who were criticized for speaking only "broken English" and living in a humble home. One writer lamented, "Every effort to persuade the parents, foreign born and steeped in centuries of superstition, failed. Even a priest, arguing long and patiently, failed to break Mrs. Vasko's stubborn determination."[55] Anna Vasko was framed as an obstacle to the treatment of her young daughter. Serving as a foil to the logical, reasonable medical experts, she was depicted as ignorant and emotional when presented with information about the "proper" care for her ill child, while the physicians who testified in the Vasko case were characterized as "philanthropic" specialists who acted "in the interests of the child" by mandating therapy.[56] The Vaskos' "otherness" and their reliance on the social welfare workers and agencies made them particularly vulnerable to judgment and new forms of state control.

On the contrary, as part of upper-middle-class urban society, the Colan family was well connected to several segments of the Chicago community and relied on the personal and professional contacts of Helaine's grandfather, Morris Hershman, during this period. For example, at the family conference, the attorney who provided legal counsel was a close friend. When the family could not reach a consensus, Hershman personally invited local physicians to serve on the expert jury. Of the religious leaders asked to review the jury's decision, one was leader of a congregation where Hershman had formerly served as president, another was the head of the Chicago Hebrew Theological College, and the third was the executive director of the United Synagogues of America who was in Chicago for a meeting. Although many experts contributed to the deliberations surrounding Helaine Colan, they were, in most cases, asked to be involved in the case and worked under the family's terms. By relying upon, not running from expert guidance, the Colans managed to retain a measure of privacy and control over their family's affairs and the management of their daughter's case.

In their simplest forms, the stories of the Vaskos, Colans, and other families of glioma babies tell a story of events that occurred alongside the child-centered institutional growth that began at Memorial Hospital for Cancer and Allied Diseases and in a new area of research around a small set of childhood cancers in the mid- to late 1930s. Professional attention grew in tandem with popular interest.

More important, however, the stories bring into relief the confluence of many factors that opened the families' personal decisions to scrutiny—factors that proved to be important in every decade of the social and cultural history of cancer. The story of cancer in the young is a story of not only the creation and dissemination of scientific and medical knowledge or professionalization and specialization but also the press' pursuit of disease-focused human interest stories, the legal system and its changing relationship to children and families, and the involvement of individual families as they intimately cared and advocated for their children with cancer from the 1930s through the present. The Vaskos and Colans provide only a starting point for a narrative of negotiation and accommodation.

"Cancer, The Child Killer"

JIMMY AND THE REDEFINITION OF A DREAD DISEASE

On the evening of May 22, 1948, Ralph Edwards, host of the popular radio program "Truth or Consequences," introduced his audience to a special guest. He announced, "Tonight we take you to a little fellow named Jimmy. We're not going to give you his last name, because he's just like thousands of other young fellas and girls in private homes and hospitals all over the country."[1] Without further explanation, the program commenced as Edwards prompted Jimmy to list his favorite Boston Braves players. Members of the team's starting lineup filed into his hospital room one by one and presented the boy with autographed baseball memorabilia. Jimmy then joined the men in singing "Take Me Out to the Ballgame" and received special permission to attend a game the next day—a day designated as "Jimmy's Day" at the ballpark. After his young guest signed off, Edwards told listeners that Jimmy was a twelve-year-old boy undergoing cancer treatment in Boston. He asked them to contribute money toward a television set for the boy's room and, more generally, to aid "Jimmy and the boys and girls of America."[2] Members of the show's audience responded generously, reportedly donating more than $200,000 to the fund and sending tens of thousand of get-well cards to Jimmy.[3] By drawing upon child-centered fundraising strategies pioneered by other voluntary health agencies, the newly created Jimmy Fund and its mission to direct research and treatment toward childhood cancers were launched with overwhelming public support.

As Memorial Hospital for Cancer and Allied Diseases's scientific and clinical activities grew and other investigators became interested in problems posed by this set of cancers, the images, voices, and stories of individual childhood cancer patients like Jimmy personalized the face of childhood cancer and helped transform childhood cancers from a set of nearly invisible diseases to a problem that attracted widespread public attention. Parents were expected to fear cancer, the "dread disease," because of its unexplainable origin, its associations with pain and suffering, and its incurability but, at the same time, maintain hope that scientific medicine would quickly develop a safe, effective cure for this deadly set of childhood diseases.

Children in Danger

The publication of *Cancer in Childhood* marked the beginning of pediatrician Harold Dargeon's child-centered work at Memorial. Although Dargeon characterized the prognosis of childhood cancer as "grave," he looked to earlier detection and record keeping as contributing important sources of knowledge that would result in better cure rates.[4] Memorial was well positioned to facilitate large-scale studies and make comparisons among children's cancer cases, but several obstacles confronted physicians charged with this task. Cancer often mimicked other childhood diseases, and without a single, definitive laboratory test it was difficult to identify suspicious cancer symptoms and make an early, accurate differential diagnosis. In the nineteenth and early twentieth century, the most common symptoms of cancer—fever, lethargy, and wasting—were so similar to the many infectious diseases that they probably remained masked. As Dargeon noted, some tumors, including those of the brain or the central nervous system, followed a "subacute or almost chronic course" and "in 50 percent of the cases of brain tumors . . . the symptoms had lasted between two and six months prior to treatment."[5] Thus, a number of children may have died with their illness undiagnosed and unreported.

Dargeon viewed systematic, comprehensive record keeping as a key arm of the cancer control effort. Conceding that childhood cancers were rare, he lamented the number of difficulties that hampered an accurate count. In the 1930s, cancer was not a reportable disease, so mortality rates could not be relied upon to determine its incidence. In the face of numerous infectious diseases at the time, public health officials did not consider cancer especially important, and the disease was neither well defined nor classified. The first hospital-based cancer registry was established in 1926 at Yale–New Haven Hospital. Dargeon warned, however, that the cases admitted to pediatric hospital wards often depended on the particular research interests of the staff. Therefore, tracking the number of children in wards or compiling autopsy records from hospitals could lead to misleading numbers. In addition, the first centralized state cancer registry (also in Connecticut) did not report data until 1935. Dargeon used mortality data from the Bureau of Records of the

Department of Health in New York City to demonstrate that, while the number of pediatric cancer deaths was not large, it was significant when compared to mortality from such childhood diseases as pertussis, congenital syphilis, cerebrospinal meningitis, measles, and scarlet fever. Statistics recorded by the U.S. Bureau of Vital Statistics in 1947 listed accidents as the primary cause of death in all children, but neoplasms followed as the second most common cause of death in children between the ages of five and nine and ten to fourteen. In the youngest cohort, ages one to four, neoplasms were listed fourth, behind accidental death, pneumonia, and congenital malformations.[6]

As part of his program on childhood cancer, Dargeon established the Children's Tumor Registry at Memorial Hospital to capture both a quantitative record of cases and a qualitative description of the common childhood tumor types. The registry attempted to more accurately measure infant and child mortality from childhood cancer.[7] Sponsored by the American Academy of Pediatrics and supervised by Dargeon, the registry's data and services reached beyond the hospital's walls to physicians and young patients across the country. The registry facilitated two goals—increasing awareness within the medical community about childhood cancer and serving as a resource for physicians. Exhibits and slides highlighting common problems in children suffering from cancer were made from the cases submitted to the registry and displayed at medical meetings and medical schools. Physicians throughout the country could consult the registry when they encountered difficult or rare questions about their own pediatric cancer cases. Physicians working in remote locations could send clinical records, tissue samples, and other laboratory data for Memorial's specialists to review. If the initial cancer diagnosis was confirmed, physicians affiliated with the registry either recommended treatment procedures or accepted the child for care at Memorial. Free treatment was provided for needy patients and families. By recording cases from Memorial's own service as well as outside sources, the tumor registry became a valuable bank of knowledge and a tool for raising awareness of pediatric cancers.

In 1939, Memorial moved and expanded on a new site donated by John D. Rockefeller, Jr. The general education board and Edward S. Harkness, an oil magnate and American philanthropist, donated funds to build and equip the new institution. Notably, the enlarged hospital facility contained the seventeen-bed Helena Woolworth McCann Children's Pavilion, which primarily served young cancer sufferers from low-income families. Approximately 300 general cancer clinics existed throughout the United States, but this was the first designed especially to house children. Articles printed in newspapers and magazines highlighted the pavilion's "children's corner," an outdoor terrace that housed a slide and carousel, a classroom designed for young patients who felt well enough to complete schoolwork, and multipurpose areas for working on crafts or singing along to the piano. The corner was compared to a family's living room with its fireplace, furniture, and toys.[8] Many photos also showed children gathered around a large, gaily painted wooden cart overflowing with balloons, plastic toys, dolls, and comic books that was regularly wheeled to the pavilion and outpatient clinic by volunteers and members of the recreation department staff. In contrast to the few photographs taken of young patients in the early 1930s that depicted bleak hospital rooms and ill children aided by slings or wheelchairs, images released by Memorial's public affairs department in the 1940s featured their expanding medical and social programs for children hospitalized for cancer.

A few years later, administrators at Memorial planned a dramatic expansion of facilities and services to increase its capacity. The major construction project that included the James Ewing Hospital, the Tower Building for Memorial Center outpatients, and the Sloan-Kettering Institute also included the McCann Children's Pavilion. To solicit funds for its annual campaigns, the public relations department used photographs of young children housed in these new spaces.

In some cases, the hospital's public affairs department masked or downplayed physical manifestations of cancer in photographs. Instead, they paired the sentimental images with provocative headlines, captions, or text to jar readers and amplify the impact of the photos. A photograph reprinted in dozens of regional newspapers featured three little boys in bib overalls walking together to the washroom toting toothbrushes, towels, and cups. In another image from the same series, one of the boys perched on a step stool to reach the sink. At first glance, such images seemed to be snapshots of healthy children caught in the midst of a simple, daily routine. The declaration "Cancer Struck All Three" and the stern warning "Guard those you love from this scourge of childhood," however, revealed the underlying message the publicity staff wanted to convey to donors.[9]

Similar techniques were also used in one of the most common types of publicity photograph—the holiday snapshot. Festivities with lavish decorations provided celebratory backdrops for many photos as children donned

Halloween masks, visited with the Easter bunny, and opened gifts from celebrities dressed as Santa Claus.[10] The caption for one such image read, "An Easter Day party in the Children's Ward of Memorial Center for Cancer and Allied Diseases. People often do not realize that cancer strikes children as well as adults. Yet more children in this country between the ages of 5 and 9 die of cancer than of any other disease."[11] The stark contrast between the joyous scene filled with spring flowers and plush toy rabbits and grim facts about a little-known disease must have shocked viewers. The use of everyday images, grave statistics, and stern warnings alerted parents of this newly recognized menace to child health but provided little information about the signs of the disease, the limited treatments, or its deadly course. Instead, the staged publicity photos focused attention on the hospital's growing pediatric spaces and services—the means for preserving and restoring child health. Photographs of patients in the children's ward were used to raise funds for Memorial Hospital generally, not only for pediatric activities. Sentimental images showing children with cancer (only a small percentage of the total cancer cases) were commonly employed by public relations departments to solicit funds for all sufferers.

According to announcements about the newly opened facilities or appeals for construction funds, there was a dire need for Memorial's initiatives. The section of the hospital's 1947–1951 quadrennial report devoted to the pediatric service claimed, "Cancer and allied diseases now constitute a major child health problem."[12] Memorial's pediatric service claimed to be the first institution in the United States to "accept responsibility for all aspects of the control of childhood neoplasms," including "demands for diagnosis, therapy, and rehabilitation of the patients, for research and for education on all lay and medical levels."[13] Despite the series of building expansion projects at Memorial and drastic increase in the number of child-focused activities in the 1930s and 1940s, the institution could not provide a bed for every ill child. Young patients from across the nation and around the world, from South America to Europe, flocked to this ward specifically dedicated to treating children with cancer. Children without reservations—even those whose families had traveled long distances—had to be refused admission. These children were diverted to other neighboring hospitals for care.

Memorial's efforts reached beyond treatment to cancer education and detection with the opening of the Strang Prevention Clinic in January 1947. The American Cancer Society's (ACS) educational campaigns had long encouraged adults to receive periodic cancer screenings. Establishing cancer centers, dispatching mobile units, operating specialized clinics within industrial settings, and supplying specialized equipment to smaller hospitals helped the ACS cancer mission reach all geographic areas. These creative means of providing cancer prevention and treatment, however, focused only on adult patients. Dargeon, Strang's first director, taught other physicians and parents about the importance of periodic physical examinations of children at home and the clinic, warning that symptoms of other, milder childhood conditions such as persistent cough, headache, vomiting, or weakness that could all be due to cancer growth. He lamented such vague symptoms were often "ignored and shrugged off as unimportant by parents." Drawing on his own experience, Dargeon also found that parents who identified frightening signs did not seek treatment earlier because they assumed that cancer was always fatal. Targeting both areas of concern, he advised his colleagues, "increased vigilance in this group should bring many children under observation in time to effect a cure."[14]

In "Children in Danger," a *Newsweek* article describing the new clinic, a photograph showed two tables of children busily working on crafts as they waited to be called for their appointments. The image promoted the idea that the young visitors to the clinic often appeared perfectly healthy but needed to undergo a complete medical assessment to ensure wellness. The clinic's medical staff reported a number of disturbing observations: "One small boy had a persistent swelling on his leg which his careless mother dismissed as a 'bump.' Another child's eyes were strangely protuberant. A third's head changed its size so rapidly he needed a new hat every few months."[15] The contradictory message conveyed by juxtaposing the image of "normal" children in the waiting room with Dargeon's warnings and the graphic descriptions of physicians' findings moved parents to action. Only two weeks after the clinic first opened its doors, appointments were filled for a year in advance.

This dim outlook toward cancer among laypersons and many physicians was warranted by a dearth of curative options, but Dargeon viewed life-prolonging therapy and experimental treatments as the best hope for achieving long-term survival. The outpatient clinic for pre- and posthospital observation dedicated part of its attention toward juvenile patients by maintaining a special pediatric clinic attended by two pediatricians and the resident. An article in *Reader's Digest* described the scene in room 102L of Memorial Hospital on Friday mornings, when children with acute leukemia returned for medical care. The head of both the hospital and scientific institute, Cornelius Packard Rhoads, expressed the depressing truth about leukemia as he observed the children playing in 102L, "We can help only 25 percent of them and they have remissions only. Their disease will recur and recur, perhaps in more violent form."[16] When asked why physicians worked so diligently to keep the children alive when they faced certain death, he responded, "We're moving faster now. Perhaps, before they exhaust their last remission, we'll have something really good."[17]

Leukemias, a set of cancers of the blood-forming tissues that affected children and, to a lesser extent, adults, merited guarded optimism from Rhoads and his colleagues. In the late nineteenth century, technological developments in microscopy had enabled physicians to observe particular blood components both qualitatively and quantitatively. New biological stains aided in the differentiation and classification of white blood cells, while hemacytometers, wells printed with a grid pattern, were used to count cells.[18] With these tools, normal blood counts and diseased states were standardized, and in 1897 the pediatric textbook *The Diseases of Infancy and Childhood*, published by L. Emmett Holt, contained color drawings of blood cells in leukemia.[19] In the twentieth century, physicians using blood counts and bone marrow aspirations definitively differentiated acute leukemia from other minor childhood illnesses. *Decreasing* platelets and hemoglobin levels alongside dramatic *increases* in leukocytes (white blood cells) signaled the presence of leukemia. Investigators found that children commonly suffered from the acute form of the disease, dying within weeks or months from uncontrollable bleeding or rampant infection.

Although Rhoads labeled the young patients "doomed," the article encouraged readers to think of the Sloan-Kettering Institute, the research facilities associated with Memorial Hospital, as a "Tower of Hope." Beyond simply diagnosing the disease, doctors were now able to briefly suspend its rapid progress in a small fraction of its patients. Many writers and parents looked to the promise of scientific research as their only source of encouragement when their child's condition and the physician's prognosis seemed bleak.

Leukemia brought Memorial Hospital and the Sloan-Kettering Institute's staff of chemists, physicists, biologists, and clinicians together to analyze a common problem.[20] At Sloan-Kettering, a research facility funded by a $4 million building grant by Alfred P. Sloan, Jr., chairman of the board of General Motors, Joseph H. Burchenal, Director of Clinical Investigation, supervised an extensive program that tested a variety of chemical compounds against transplanted cancers in mice and other animal models. Each week, dozens of chemotherapeutic agents arrived at Sloan-Kettering from commercial laboratories, pharmaceutical companies, university scientists, and physicians. Promising results produced at Sloan-Kettering were then applied clinically at Memorial.[21] The *Reader's Digest* article labeled Memorial Hospital the "Human Laboratory," yet noted, "the patients in Memorial are never used as experimental subjects. But virtually all patients beyond the help of surgery are willing to have new treatments tried on them."[22] As at other research institutions, the line between patient and subject was blurred.

In the late 1940s, Memorial Hospital became a center for medical training opportunities in childhood cancer at the graduate and postgraduate levels. In 1948, the American Board of Pediatrics approved the Memorial Hospital pediatric residency to educate pediatricians in the diagnosis and management of tumors in children. Other pediatricians completed shorter rotations through the pediatric service at Memorial Hospital or attended lectures or courses given a few days a week on the problems of childhood neoplasms. Medical students received a portion of their training at Memorial and then disseminated their knowledge during their residency or practice. Thus, some children turned away from Memorial and sent to other area hospitals still benefited from the institution's programs. Despite this new, advanced training program and the other child-centered activities pioneered at Memorial in the 1940s, the public soon became aware that a diagnosis of acute leukemia or other common childhood cancers equaled a certain death sentence for children.

Cancer, The Child Killer

Cancer had long inspired alarm among laypersons because of its associations with pain, disfigurement, and inevitable death. In 1913, the American Society for the Control of Cancer—renamed the American Cancer Society in 1944—began to lead public education efforts about the dread disease. At the time of the organization's inception and during its early years, physicians and surgeons dominated its small membership and conducted much of the public education program, initiatives that narrowly targeted an adult audience. Beginning in 1936, the Women's Field Army of the ASCC set out to educate the public about the seven warning signals of cancer. Focusing on prevention and early detection, their door-to-door campaigns, radio spots, printed materials for industrial sites, and articles in popular magazines all urged Americans to raise their awareness and to seek prompt treatment by regular physicians if they observed any suspicious signs.[23] Diagrams in the organization's pamphlets pinpointed common cancer targets on simple line drawings of mature male and female bodies, highlighting tumors of the breast, lungs, and reproductive organs.

While the ASCC narrowly targeted cancer in adults at the organization's inception and throughout its early years, it used new methods to raise awareness of cancer in America beginning in the 1940s. As James Patterson has argued, the ASCC owed much to the leadership of Mary Lasker, a society woman and health philanthropist, for this transformation.[24] Under Lasker's direction, the newly renamed ACS emulated the March of Dimes campaigns led by the National Foundation for Infantile Paralysis, implementing modern advertising concepts to raise funds and spread their messages of cancer prevention and control. The ACS campaigns encouraged Americans to not only not fear cancer but also actively fight the "dread disease" on both a national scale and personal level. Focusing on prevention and early detection, Americans were urged to seek prompt treatment from orthodox physicians if they observed any suspicious signs. The ACS and its health writers gradually applied this message to cancer in children.

At first, health writers inserted brief sections about childhood cancer into their general cancer articles. *Hygeia,* a magazine published by the American Medical Association for lay readers, featured a monthly series on cancer awareness and prevention. The author of one article asked his readers, "Do You Fear Cancer?" and recommended the proper attitude, information, and actions to adopt when confronted with the disease. Its messages mirrored those promoted by the ACS: the problems caused by delaying diagnosis and treatment, the value of regular exams by orthodox physicians, and the dangers of "quack" healers. Only one sentence addressed childhood cancer: "A misconception which has proved costly is the belief that cancer occurs only in older patients. This theory has long been discarded, for cancer does appear in steadily growing numbers among young people."[25] By the end of the decade, a few full-length pieces on childhood cancers dotted the pages of popular magazines.[26]

As the number of articles about cancer increased, authors (sometimes physicians) began to enlarge their scope to include children as possible cancer victims. Popular articles in women's magazines began identifying cancer as a menace to children, proclaiming, "Cancer Kills Children Too!" and "Cancer, The Child Killer."[27] Authors alerted parents—especially mothers—to cancer's threat and entreated them to minimize their risk by watching for common cancer symptoms and scheduling regular physical examinations for their child. In the March 1947 issue of *Woman's Home Companion,* Frank Rector, secretary of the Cancer Control Committee of the Michigan State Medical Society, provided basic information about cancer in children and classified the common cancers according to the most frequent age group affected by each type. He reported that Wilms's tumor (kidney), eye tumors, and central nervous system tumors struck the youngest age group, brain tumors and leukemia targeted those between the ages of five and ten, and bone tumors affected those older than age ten. Most authors, by contrast, combined factual information with distressing headlines or frightening testimonials in order to capture readers' attention and rally them around this newly identified set of diseases. Articles advised mothers to work with their child's pediatrician to identify suspicious signs of cancer and procure timely treatment for the young. One physician wrote,

> The best hope for preventing needless deaths lies with mothers everywhere who must recognize that there is such a thing as cancer in children and that it isn't something to be frightened of or fatalistic about. The signs of cancer in children are so often those which any observant mother cannot help but notice.[28]

Because cancer's symptoms often mimicked those of common childhood ailments, such advice may have needlessly induced unwarranted anxiety and blame. Cancer joined a long list of conditions for which mothers had responsibility.

A Grave Burden

Cancer placed a grave burden on mothers—for the timely diagnosis of their children's illnesses, the losses of their children, and the societal blame attached to all phases of the disease. Since the late nineteenth century, authority over children's health in America had been a contested subject between domestic caretakers and medical professionals. Whereas lay experience equaled expertise early in the period, the rise of the scientific, formal health system later challenged women's partnership in medical knowledge and practice.[29] Pediatricians became proponents of a new bond between mothers and specialists in child health by integrating child-rearing wisdom with professional practices to formulate methods of scientific motherhood.[30] Experts endorsed regimented feeding and sanitation techniques to prevent infectious disease and reduce mortality, a considerable responsibility for all mothers.[31] In the 1930s, the range of experts again expanded as psychologists and child guidance experts began to advise women in the proper way to discipline and feed their child and to preserve their child's mental and emotional health.[32]

Molly Ladd-Taylor and Lauri Umansky have argued that mother blaming reached its peak during World War II and the postwar years when mothers were targeted for seeking wage work, contributing to the decline of the home, and creating juvenile delinquents.[33] Mothers were expected to rely on expert guidance as a corrective. Published in 1946, Benjamin Spock's bestseller *Infant and Child Care* addressed physical and emotional health by providing detailed descriptions of specific methods appropriate for each developmental stage and championing a permissive style of childrearing; the pediatrician's readable manual became an essential handbook for many mothers. Mothers also retained responsibility for maintaining child health, though the nature of childhood diseases shifted from acute to chronic conditions. Ladd-Taylor and Umansky noted, "Death serves as 'evidence' of failed motherhood; the refusal (or inability) to protect one's children from danger, or even from disease."[34] Mothers who neglected to follow professional guidance at home were blamed for the poor health or deaths of their children.

By employing rhetoric of maternal responsibility in childhood cancer articles, authors blamed mothers for their child's death during a period when a timely diagnosis generally did not contribute to a better outcome for childhood cancer victims. In the 1940s, few effective treatments beyond surgery and radiation were available for childhood tumors and a diagnosis of acute leukemia was an inevitable death sentence; however, mothers' frequent interactions with their children meant that medical professionals relied heavily on them to make vital decisions about the seriousness of symptoms and the appropriate action to take after detecting any suspicious signs of cancer.

During World War II, reports about special furloughs granted to servicemen whose children were gravely ill with leukemia or other cancers appeared in the popular press.[35] The stories reinforced a turn toward domestic life that had occurred as the hard times of the Depression gave way to a strong wartime economy. Birth rates rose from a low of 18.4 per 1,000 women during the Depression toward a high of 25.3 per 1,000 in 1957, and the average family size grew markedly during this period.[36] Children were viewed as the center of the patriotic American family and a source of security during a tumultuous time. As husbands and fathers left to serve overseas, wives were entrusted with responsibility for managing the home, securing jobs in war industries, and, most important, rearing children. In a few publicized cases, though, families separated by war were reunited prematurely under the dark cloud of childhood cancer.

The travails of the Truax family made the *New York Times.* In 1944, the mother of eighteen-month-old Therese Truax sent a letter to President Roosevelt requesting that her husband be returned from service in the South Pacific to visit their dying child.[37] Therese had been diagnosed with acute leukemia and had little time left. Mrs. Truax rushed her baby from the local New Jersey hospital to Babies Hospital at the Columbia-Presbyterian Medical Center desperately hoping to find a new treatment or cure, but the doctors had no lifesaving treatments to offer. Therese's father received a month-long leave via presidential order and immediately flew back to the United States via military plane, and doctors gave her regular blood transfusions in order to prolong her life until her father arrived home. Upon his return to New Jersey, the newspaper reported that he shared his wife's view that there was "real hope" for the recovery of the infant, but only six days after his well-publicized return, baby Therese died of leukemia and related complications.[38]

A cluster of stories in newspapers and magazines told of other urgent wartime reunions. A young mother in the Midwest found a mysterious swelling on her nine-month-old son Arthur's groin. After a week of observing little change, she took him to the family physician. The doctor excised a tissue sample and sent it to the state capital for careful study. When the second opinion confirmed the initial cancer diagnosis, Arthur and his mother traveled to Memorial Hospital in Manhattan. Arthur's father, who was serving in Frankfurt, Germany, received a cable containing urgent news of his son's precarious condition. When he arrived in New York to join his wife and to see his son for the first time, the swelling had been successfully removed and Arthur was recovering. His father proclaimed, "Thank God my wife acted so quickly!" to interviewers reporting the story of his son's illness and his own return from serving overseas.[39] In 1946, Lieutenant Keith DuBois flew from Germany to Memorial Hospital on emergency leave to see his eight-month-old son, Allan, for the first time. His wife and son had come to New York from Green Bay, Wisconsin, after his pediatrician had recommended the trip. His father affirmed that it was "wonderful" to see his son but when questioned further about his homecoming admitted, "This isn't the way I dreamed about it . . . this isn't the way I wanted it."[40] A fourth father rushed home from his post at Pearl Harbor to be at his dying daughter's bedside.

These news stories emphasized the consequences of prompt cancer detection by watchful mothers. Mothers could save or prolong their child's life by detection and proper treatment or, at the very least, could arrange a final family reunion during a child's illness or death. The articles also repeated themes of wartime heroism and underscored the importance of family togetherness during a tenuous time. By granting emergency leave and providing transportation directly to hospital sites, the federal government and military administration illustrated value of children and family at the national level.

Some mothers immediately turned to their local physicians and then nationally known specialists for help, but others (like the Vaskos and Colans) intentionally delayed treatment. In "Cancer, The Child Killer," Lawrence Galton noted that this attitude occurred more frequently with cancer than other diseases. "Doctor after doctor has his hands tied when parents refuse to let him take a tiny section of tissue from the suspected area and analyze it to make certain."[41] This denial may have been due to several factors, including fear and dread of a cancer diagnosis, ignorance about the disease, a lack of available medical care, and, at times, thoughtful resistance to the proposed treatment. Articles about such families used them as examples that even a short delay could mean death for a child with a benign or malignant cancer that hindered his or her growth or impeded a vital physiological function. At times, they also served as thinly veiled critiques of immigrant families, those living in rural areas, or families of low socioeconomic status.[42] The articles implied that all parents (especially mothers) were to act quickly and should follow the advice of physicians or, ideally, cancer specialists without hesitation.

Dying Young

The story of sixteen-year-old Robert de Villiers' death from acute leukemia in 1944 poignantly demonstrated that no physician could slow the progression of the disease, let alone provide a cure. Parents, nevertheless, maintained hope in the face of terminal prognoses. "We had hoped," Robert's father recorded in his diary, "that we would succeed in keeping our Robbie alive until a cure had been found."[43] Only three months after physicians made their initial diagnosis, Robert died. Robert's unpreventable death from acute leukemia demonstrated the disease's rapid course and fatal outcome in children. In the early 1940s, no life-prolonging treatments were known, but physicians used radiation therapy to reduce pain and swelling in children's joints and ordered blood transfusions to strengthen young leukemic patients, albeit temporarily. Children diagnosed with acute leukemia at Memorial in the early

1940s survived an average of only 27 days following admission.[44] Severe side effects, including infections and massive hemorrhages, commonly caused death rather than the disease itself.

A few physicians cautioned their colleagues and a general readership that the proliferation of the disease could not be stopped by medical means. In 1946, William Dameshek, a hematologist at Memorial, was outspoken in challenging the optimistic portrayal of acute leukemia and false hope furthered by the popular media. Addressing these misconceptions, he wrote,

> There is nothing worse than to make a diagnosis of *acute* leukemia. The tabloid press, once it hears of a case, gives all the harrowing details, and the radio blares loudly of blood donors, particularly from "cured" cases, and of the new "atomic bomb" treatment and other matters. The family is deluged with suggestions, and in the meantime the patient goes steadily downhill. Much as one hates to admit it, there is practically *nothing* to offer in the acute or rapid case of leukemia.[45]

As a way to describe leukemia to *Hygeia*'s lay readership, he compared a person with leukemia to a healthy garden overtaken by "an overgrowth, like a weed, of a special type of white cell that arises in one of the three white cell forming tissues."[46] Dameshek warned that particular challenges faced the few physicians and scientists who dared to devote themselves to the investigation of this disease: difficulty in diagnosis, unknown etiology, and no efficacious therapies. Despite his grim assessment of the available treatment options, Dameshek alluded to a series of highly secret series conducted through the war on the effects of mustard gas and related substances on the blood. Although mustard gas caused only temporary relief for patients with the acute form of leukemia, some scientists had begun researching related agents as wartime projects and equipment were redirected to investigating and improving civilian health after the war had ended. Dameshek concluded the article on a positive, yet tempered, note, hoping for a scientific breakthrough on their research related to the "leukemia monster."[47]

The de Villiers, devastated by the loss of their son and frustrated at the inability of physicians to cure him, established the de Villiers Foundation in 1949 to memorialize their son and support leukemia research in the hope that a cure could be found for others. The foundation, headed by Robert's mother, Antoinette de Villiers, began in a Manhattan office with two thousand dollars and a small staff of volunteers. Following public announcements about possible medical applications of chemical warfare agents, the foundation elected to support this proliferating, multidisciplinary field of research. This organization, renamed the Leukemia Society of America in 1955, sponsored annual, international research competitions with monetary awards seeking a treatment for the devastating disease.[48] At a time when there was little government support for research in childhood cancers, voluntary agencies helped to fill this role.

The Leukemia Society also coordinated fundraising and public education efforts. During its early history, from 1949 to 1955, many of the contributions were given in honor of family members or friends who had died of cancer. The Leukemia Society also acted as an information clearinghouse by releasing statistics and news on recent research developments to the press. Public education was the third branch of the Society's activities. The group published brochures that described leukemia and furthered the need for intensifying lines of research directed toward the disease, they broadcast a panel discussion on the Society's major functions over local radio and television stations, and Allen Funt, the producer of the television program *Candid Camera,* made a documentary film on the subject.[49]

Jimmy Captures the Limelight

As the de Villiers family established their foundation, the first chemotherapeutic agent effective against acute leukemia was evaluated in the lab and clinic. Sidney Farber, chief pathologist at Children's Hospital in Boston, observed that daily intermuscular injections of aminopterin (4-aminopteroylglutamic acid), a folic acid antagonist that disrupted cancer cells' metabolism—halting all cell activities, including growth and reproduction—induced temporary remissions in ten of fifteen young patients with acute leukemia.[50] When Farber first publicly announced his findings he asserted, "It is the most wonderful hope we have, and we know now that with this drug, and with other chemical agents, we are working in the right direction."[51] Previous experimental therapies had not produced an effect in the 300 patients with the disease treated at the hospital.

Farber had collaborated with Yellapragada Subbarow of Lederle Laboratories, a unit of the American Cyanamid Company in Pearl River, New York, to test the effects of folic acid and other compounds on disease.[52] Correspondence between Farber and Subbarow in 1944 alluded to Farber's interest in folic acid. In March 1944, Farber wrote asking Subbarow about the general contents of capsules containing liver powder and, "in particular, approximately how much folic acid these capsules contain."[53] Louis Diamond, a pediatric hematologist at Boston Children's Hospital, had observed promising results after administering a particular liver fraction on patients with several different types of anemia, especially a severe form. On hearing these results, Farber requested an additional supply of capsules for his laboratory animals, pediatric cancer patients, and older patients he was trying to study in neighboring hospitals.[54] Farber also collaborated with other physicians in the Boston area. George Foley, first of the Department of Preventative Medicine at Harvard Medical School and then of Massachusetts General Hospital, conducted preliminary laboratory studies to evaluate folic acid as a possible treatment for acute leukemia. On April 8, 1948, Farber made his public announcement that a drug had induced temporary remissions in acute leukemia patients.

Newspaper reports of the announcement minimized the collaborative nature of the folic acid investigations, but an extended letter sent to Subbarow revealed Farber's view of the events, his reliance on related scientific fields, and the importance of joint programs between industry and academia. In his letter, Farber amended the articles and emphasized the valuable contributions made by Lederle Laboratories. Farber characterized the overall research program as a cooperative one, stating,

> We in biology and medicine can make no progress without the cooperation and research of men in the various branches of chemistry; those in chemistry require the biologic proof . . . there is nothing but good that can come from such association of scientists who approach the same problem by different routes.[55]

He assured Subbarow and his associates that proper credit would be given in his published work, including future publications. He also invited representatives from Lederle to participate in the First Conference on Folic Acid Antagonists in the Treatment of Leukemia, a symposium sponsored by the institutions involved in the aminopterin research program including Harvard Medical School, Children's Hospital, Peter Bent Brigham, and Deaconness Hospital held in January 1949.[56]

Farber and many science writers advanced the discovery of aminopterin as a preliminary step in the synthesis and testing of chemical agents for all systemic cancers and solid tumors, but his results invited skepticism among some clinical investigators who doubted that Farber had been completely truthful when describing his experimental design, methods, and results. Doubts grew when researchers were unable to reproduce Farber's findings in adults with cancer.[57] Dr. Leo Meyer, a physician testing aminopterin in older patients, met with Farber's group to discuss his data. Farber wrote Subbarow to share the results of the discussion:

> It appears that Dr. Meyer has not had an encouraging experience with aminopterin in his adult patients. Our experience with children so far in acute leukemia has been striking from the point of view of clinical improvement and important hematological changes. We have never used more than 1 mg. a day and are using as little as 1/2 mg. a day.[58]

Confident of his results, Farber and his group began administering aminopterin and a closely related drug in sequence to test whether two drugs would induce a longer remission in a pediatric patient population.

As a way to sidestep his critics, Farber looked outside his home institution for endorsement of his work, becoming a leader and spokesperson for the ACS, providing expert testimony at congressional hearings, and establishing a new fundraising organization. By forging partnerships with civically minded groups in the Boston area, Farber established the Children's Cancer Research Foundation (CCRF), a regional platform for expanding the provision of advanced treatment for children with cancer and furthering his own clinical program.

The initiative began with a major commitment from the Variety Club of New England. Variety Club International, an organization of men in the motion picture and theater business, first formed from a philanthropic project focused on an abandoned child. After finding a child in a Pittsburgh movie theater in December 1929, a group of showmen devoted themselves to the child's well being. In the years that followed, the organization's efforts branched out in many directions, but remained focused on the spiritual, physical, and medical needs of underprivileged children. In 1947, the Variety Club pledged $50,000 to establish a blood bank and blood research department at the Children's Medical Center in Boston. After one of the club's committees toured the children's cancer ward during a hospital visit and learned about Farber's promising investigations, they joined with the Boston Braves to establish the CCRF. The Massachusetts Chiefs of Police Association also later adopted the foundation as its official charity.

Two years after the March of Dimes introduced its first "poster child," the CCRF began planning for its inaugural event. It molded its young representative by changing the boy's name from Einar Gustafson to "Jimmy" to protect his privacy and, perhaps, to attach a popular, favored boy's name to their efforts. By choosing this pseudonym and playing up his avid interest in the local professional baseball team, publicists created a poster child with all-American attributes and interests. Jimmy served several important purposes: he appealed to potential donors, personified cancer, and reminded listeners that cancer did not spare children. The Variety Club launched the Jimmy Fund in 1948 by hosting Edwards's national radio broadcast from his bedside as a way to build an intimate connection between a patient and audience members. The original Jimmy only participated in the fund's launch before returning to his family's farm in Maine, but the foundation retained its focus on child sufferers by permanently associating his name and a sketch of a boy's profile with all of its fundraising and promotional activities.

The Jimmy Fund employed a number of visual techniques to convey their message that, by contributing dollars to biomedical research, a cure for cancer would surely be discovered. Fundraising canisters placed at stores and sporting events used a "before and after" strategy also pioneered by the March of Dimes. The fund pictured a line drawing of a boy in a wheelchair gazing out a window under the line, " I can DREAM can't I?" Beneath the plea, a second image displayed the boys' wish to slide into home base, barely evading the catcher's efforts to tag him. Overhead, the umpire declared him "safe." Between the two images, the legend "Jimmy Fund" was boldly printed to identify the boy's source of hope and recovery. Fundraising canisters, movie theater collections, ballpark promotions, and appearances by such celebrities as famed Red Sox outfielder Ted Williams prominently featured children in their appeals.[59]

Early donations to the CCRF financed the construction of the Jimmy Fund Building, a facility nicknamed the "Building of Hope" and now part of the well-known Dana Farber Cancer Institute. At the formal dedication day ceremonies on January 7, 1952, Farber and other noted speakers reaffirmed their commitment to a child-centered fundraising mission, research agenda, and building design. Farber said that he mobilized his staff by framing research projects around "a given patient—a patient with a name, a patient with a personality, a patient, the child of parents who are concerned over the welfare of their child."[60] The individual patient who suffered from a particular cancer—like Jimmy—became the motivation for his research programs, and the "total care" model implemented at the facility. This model emphasized the physical, emotional, and social components of disease as they affected patients and their family members. Similarly, the comments made by the dean of the Harvard Medical School (an affiliated facility) underscored the particular tragedy caused by malignant disease when it affected a young person. He posited that

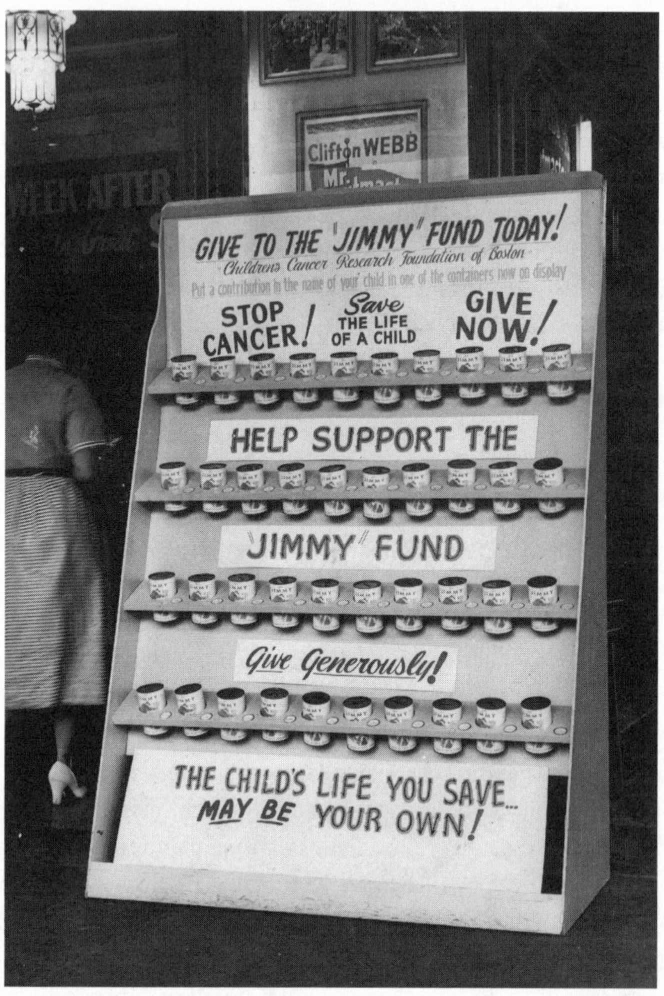

The advent of coin-collection canisters helped the Jimmy Fund raise more than $200,000 by the end of its first campaign season. Canisters—often labeled with the names of young patients with cancer—were commonly displayed at the entrance to ball games or were passed through movie theaters by ushers (above). Poignant film clips featuring Hollywood stars such as James Cagney, Joan Crawford, and Spencer Tracy urged patrons to give generously (opposite). Reprinted with permission of the Dana Farber Cancer Institute.

it was the age of the victims that had moved donors to give generously. A third speaker framed the link between the Jimmy Fund and the protection of children in terms of national strength and pride, stating that the people of New England had generously undertaken "this project dedicated to the alleviation

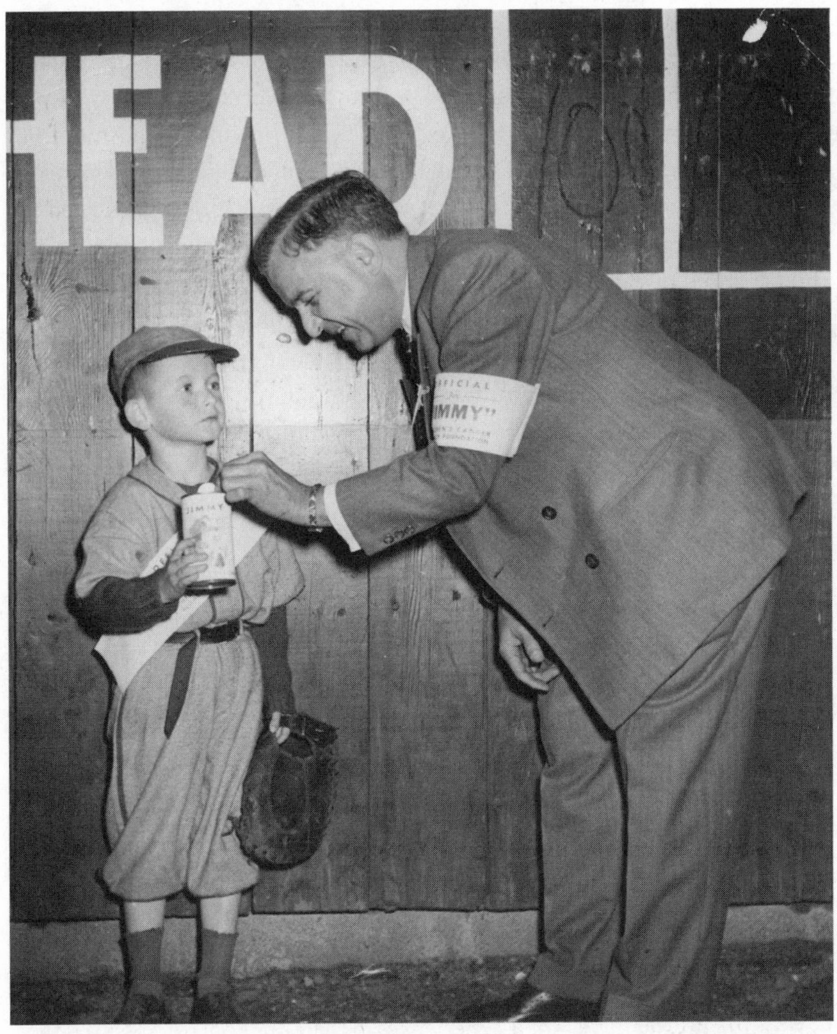

of suffering among children and the building of strong bodies and strong minds in true American tradition."[61] Nearly a dozen prominent physicians and administrators of medical research facilities—including J. R. Heller, director of the National Cancer Institute; Martin J. Mullin, president of the Children's Cancer Research Foundation; Shields Warren, director of the Division of Biology and Medicine of the Atomic Energy Commission; Samuel Pinanski, chief barker of the Variety Club of New England; and Jim Britt, a national sportscaster and trustee of the Children's Cancer Research Foundation—emphasized this two-part message common in the 1940s: children waged courageous, yet futile battles against deadly cancers, but targeted scientific research would insure medical progress and certain cures.

The speakers' comments also revealed the value of children in America—as individuals, as part of a defined cohort, as family members, and as future citizens. Premature death from cancer was a tragedy that deeply affected the family as well as the prospering nation. Rather than simply appending new child-focused research and treatment programs to an existing cancer center, pediatric cancer was now promoted in Boston and in New England as a major problem of child health that merited the public's support and an independent facility. For many, Jimmy symbolized the tragic, premature loss of innocent lives but also a source of hope that a cure was near.

In the late 1930s and 1940s, as mortality from infectious diseases of children declined, the definition of cancer expanded from a dread disease common in adults to a rare, but deadly threat to children's health. During the 1940s, Memorial created new spaces, services, and training programs that addressed the needs of a new population, and preliminary results from clinical chemotherapeutic research demonstrated that a chemical agent could temporarily halt acute leukemia. Fears heightened about the silent growth and expanding reach of cancer in children, but hope about the promise of new, curative chemical agents characterized medical and popular literature.

Stories of individual sufferers emphasized to the public that cancer potentially threatened the life of every young person and the happiness of every family. By founding their organizations in reaction to the illness or death of specific cancer victims, the Leukemia Society of America and the Jimmy Fund personalized the public face of cancer in children. They prioritized the child by prominently featuring their experiences in their materials but assured their audience that death from cancer was not inevitable. At a time of concern over the unpredictability and paralytic power of polio and great enthusiasm about the promise of scientific research, cancer organizations, physicians, and parents of childhood cancer sufferers often embraced and promoted an optimistic message like the one spread by the March of Dimes—that debilitating disease could be conquered by supporting scientific and medical research. But what were the emotions and obstacles experienced by young patients with cancer and their caregivers at this time? *Death Be Not Proud,* a slim volume that documented one noted family's yearlong struggle with cancer, illustrated the turmoil that accompanied the tragic, premature loss of innocent life.

CHAPTER THREE

Death Be Not Proud

CHILDREN, FAMILIES, AND CANCER IN POSTWAR AMERICA

In 1947, Johnny Gunther, a seventeen-year-old boy, died of cancer. *Death Be Not Proud,* a best-selling memoir written by his parents, recorded his fifteen-month fight against a fast-growing brain tumor and eventual death. Excerpts in the *Ladies' Home Journal* and *Reader's Digest* gave readers a glimpse into the medical management of children's cancers in the 1940s. When medical specialists assigned to Johnny Gunther's case admitted to knowing little about the boy's rare brain tumor, its course, and effective therapies, his father, John, and his mother, Frances, carried out a relentless and ultimately fruitless search for more information and new treatments. Hoping to find a cure for their teenage son's rare cancer, the Gunthers consulted more than thirty physicians in their desperate quest. In *Death Be Not Proud,* his story and his parents' experiences reached millions of Americans, and private illness entered the public realm. In the late 1940s, Johnny's story joined that of "Jimmy" and other children in alerting parents that cancer was no longer solely an adult disease but rather was one that threatened people of all ages. Shockingly, their own child could be struck by cancer.

Death Be Not Proud had its most profound effect through its content and wide readership. In the book's opening pages, John Gunther described the book as "a story of what happened to Johnny's brain."[1] Yet, this was only one thread of the narrative. Gunther's portion of the book provided a detailed chronological record of his son's illness. Gunther documented Johnny's diagnosis, the rapid progression of the disease, Johnny's repeated hospitalizations and procedures, and the family's search for effective cancer therapies but focused on Johnny's protracted struggle to continue living while hovering so close to death. "I write it," Gunther explained, "because many children are afflicted by disease, though few ever have to endure what Johnny had, and perhaps they and their parents may derive some modicum of succor from the unflinching fortitude and detachment with which he rode through his ordeal to the end."[2] Gunther's story told of a "long, courageous struggle between a child and Death" by firmly placing his son at the center of the narrative. He quoted Johnny frequently and described his daily activities and accomplishments in detail. By including selected entries from Johnny's diary and excerpts of letters written to his parents, Gunther further established his son as the book's protagonist and authorized his voice.

Death Be Not Proud and the overwhelming reader response that followed demonstrated how the plight of one boy could influence the public's awareness of cancer's menace to all ages, but the Gunthers' poignant story had a larger impact. At a time when Americans embraced the promise of conquering disease though scientific and medical methods and championed the child-centered family, letters sent to the authors of *Death Be Not Proud* suggested that incurable diseases, especially those that affected children, threatened these postwar ideals. The memoir also created an unprecedented forum for parents to share publicly their personal feelings about rarely discussed topics—illness, loss, and grief—especially with regard to their children. Through their correspondence, parents exchanged advice, regained hope, and found solace in their common experiences.

John Gunther

Death Be Not Proud received widespread attention because of the literary reputation of its primary author, John Gunther. A Chicago native, Gunther graduated from the University of Chicago in 1922 and began working as a foreign correspondent for the *Chicago Daily News.* Reporting on events in Central Europe, he found he preferred to provide "human rather than purely political portraits" of current events.[3] Cass Canfield, an editor at Harper and Brothers, persuaded Gunther to apply this unique approach to an extended examination of the European continent.[4] Gunther traveled rapidly through urban and rural areas, interviewing dignitaries and common citizens along his route. In 1936, Gunther published *Inside Europe,* the first of his best-selling "Inside" series that described the economic and political climates of various countries and continents. Following the success of this volume, he published *Inside Asia* (1939) and *Inside Latin America* (1941). Although some critics dismissed his treatments as "superficial," his work was praised by reviewers and widely translated.[5] Gunther expanded his readership and funded the travel required to research the series by publishing short book excerpts in *Reader's Digest* and other popular magazines such as *Look* and *Collier's.*

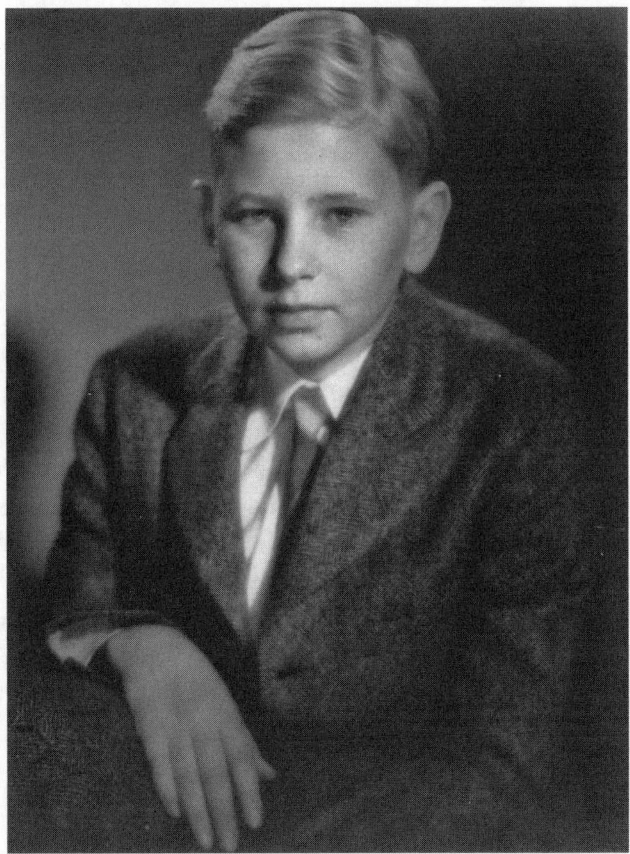

Johnny Gunther, *Death Be Not Proud.* This formal portrait, taken two years before Johnny's illness, faces the title page of *Death Be Not Proud.* While the book is a detailed retelling of the Gunthers' experience with cancer, the placement of this photograph identified Johnny as the center figure on whom the memoir was based. Reprinted with permission of the Schlesinger Library, Radcliffe Institute, Harvard University.

In the years 1944 to 1947, Gunther researched and wrote *Inside USA*. Employing a research method similar to that of his previous books, he undertook a state-by-state survey of the country, conducting approximately a dozen interviews each day. Because of the popularity of his earlier work, newspaper writers in the Midwest and South reported on Gunther's activities and his presence in several towns was printed as front-page news in the advent of *Inside USA*'s publication in March 1947. Gunther recalled, "Harper's advance sale was, I was told, the largest in its history for a trade book, and one piquant detail is that Macy's in New York did 90 percent of its book business with this single title on one day shortly after publication."[6] The book topped the best-seller list by the summer and became one of the best-selling works of nonfiction in the history of American publishing up to that time.[7] Gunther had worked months past his publisher's deadline to complete the book by March 1947. In his short autobiography, he recalled, "Most of the writing had to be done under the pressure of acutely difficult and painful circumstances: my son Johnny's long illness."[8]

In *Death Be Not Proud,* Gunther chronicled the events of his son's prolonged illness. The slim volume resembled his "Inside" series in that it revealed the in-depth research Gunther single-mindedly pursued to find a treatment or cure for his son. The book's title, the first words of a holy sonnet composed by the English poet John Donne, suggested his motivation for publicly sharing these personal events. Donne's poem (c. 1610) robbed death of its power, concluding, "One short sleepe past, wee wake eternally, / And death shall be no more; / Death, thou shalt die."[9] Through *Death Be Not Proud,* Gunther hoped to lessen death's grip over himself, his family, and other cancer sufferers. In the book, he immortalized his son and contributed to cancer research by raising awareness of the disease and by donating the publisher's profits and his own royalties to cancer research.[10]

"A Word from Frances"

John Gunther's name appeared on the title page, but *Death Be Not Proud* was a collaboration between John and Frances Gunther, his ex-wife and Johnny's mother. A short section titled, "A Word from Frances" followed John's detailed chronology in which she contributed a personal view of Johnny's illness and her family. John and Frances had married in 1927 and Frances joined her husband in the professional world of journalism and politics. She assisted John with the first two "Inside" books, covered Central Europe and the Balkans for the *London News Chronicle* from 1930 to 1935, and published a collection of writings about India based upon extensive travels in the Middle East and Asia. Amid her professional career, Frances became pregnant with their first child, Judy. Before she turned one, Judy died from a condition called "thymic death." After the sudden loss of her child, Frances sought professional advice from L. Emmet Holt, an eminent pediatrician at Johns Hopkins Medical School. Holt replied that physicians knew little about the cause or nature of thymic death.[11] In 1930, the Gunthers' second child,

The Gunther family. Family correspondence reveals that John Gunther was initially critical of Frances's contribution to the book, but their respective sections—and Johnny's words—demonstrated that each had a strong relationship with their son and made distinct imprints on his life. Readers' letters applauded the parents' individual efforts to provide the best care for Johnny in his final months. Reprinted with permission of the Schlesinger Library, Radcliffe Institute, Harvard University.

Johnny, was born while the couple was living and working in Paris. Although the couple divorced in 1944, they shared responsibility for Johnny's care before and during his illness.

In her section, Frances, like John, selectively documented aspects of her son's last months. Complementing John's section on their son's medical care and his prolonged hospitalizations, Frances recounted the time she and Johnny had spent together at her home in Madison, Connecticut. She fondly recalled her son's zest for reading, attempting chemistry experiments, identifying and collecting rocks, and discussing philosophy and religion. Much of Frances's essay addressed Johnny's death and her profound sense of loss that followed. Frances's private papers recorded her sorrow. On New Year's Eve, six months after Johnny's death, she wrote, "Johnny'd be 18 and nearly 2 months and home for Xmas holidays from Harvard and out dancing tonight—and sleep late in the morning—and get up at noon, fuzzy and towering in his bathrobe—and I'd hug him—Dear God, how I'd love to hug you, Darling."[12] In May, a diary entry revealed Frances's intense feelings of self-blame for her children's untimely deaths. She wrote, "Oh, I let Johnny die and I let Judy die—your great gifts to me, O God, I let them die—How could I let them die?"[13] Although Frances concealed such private writings from public view, her essay shared her intimate anxieties about the responsibilities of parenthood and the anguish of losing a child. She directly addressed other parents, both those who still had their children and those who had lost children to illness or war.

In private notes, John Gunther called Frances's section "beautifully, movingly, exquisitely written," but he also criticized its contents.[14] He summarized the two sections: "I tell a story and she tells of a relationship and how you stand it when a relationship is broken by something external."[15] He opined, "It's so much by a *woman.* Women *are* different from men. It's so *warm.* But mine is *strong, solid.*"[16] Despite Gunther's complaints that Frances's section was too personal to be published, the publishers included it in the volume for its added perspective on Johnny's illness.

Johnny

In constructing *Death Be Not Proud,* Gunther relied on his copious notes and his son's letters and diary to make Johnny's narrative an important part of the memoir. These sources enabled him to insert Johnny's lively dialogue throughout the account. "I am quite a guinea pig!" exclaimed Johnny when faced with the initial onslaught of diagnostic tests and medical procedures.[17] Gunther also included Johnny's constant demands to be told about the dire nature of his condition and prognosis despite his physicians', nurses', and parents' attempts to hide the truth.[18] Frances's words animated him in readers' minds. She wrote,

> An open fire with a broiling steak, a pancake tossed in the air, fresh nectarines, black-red cherries—the science columns in the papers and magazines, the fascinating new technical developments—the Berkshire music festival coming in over the air, as we lay in the moonlight on our wide open beach, listening—how he loved all these![19]

Johnny's diary entries and letters also contributed to the reader's view of Johnny. Excerpts from his brief diary entries informed the reader about his daily activities. His letters to friends and teachers revealed his concerns about his schoolwork and his progressing illness. Johnny's voice and experiences, like those of Jimmy, Robert deVilliers, and others, alerted a broad audience of cancer's effects on young sufferers. The inclusion of Johnny's experiences also makes *Death Be Not Proud* an exceptional source for an historical examination of a child's experience of illness.

Historians have explored topics related to children and childhood, child health, and childhood diseases, but few have captured the experiences of children themselves.[20] A literary genre commonly labeled illness narratives has aided historians in examining the patient perspective of those suffering from childhood diseases.[21] Through autobiographical or biographical accounts, patients document and derive meaning from their experiences of illness, treatment, or recovery and rehabilitation. Illness narratives have been written about a variety of diseases in children, including hemophilia, mental illnesses, and polio.

Several scholars have used illness narratives to supplement institutional records and to study and compare polio sufferers' experiences.[22] Rather than focusing on national events such as the March of Dimes, published polio narratives, interview transcripts, and letters sent to President Roosevelt reveal sufferers' experiences of illness and rehabilitation. From these sources, patients' own words help provide insight into patients' daily physical and psychological struggles, their family's concerns, and their hope for recovery. In listening to children's words, Kathryn Black found that their explanation of their experiences differed significantly from the descriptions recorded by their adult caregivers.[23]

Like polio narratives, *Death Be Not Proud* is a valuable historical tool. It serves as a starting point for a study of childhood cancer narratives published from 1949 though the present, a narrow window into an examination of the postwar child and family, and a way to gain insight into one patient's and family's experience of illness in the late 1940s—a time when cancer in children began garnering the attention of physicians, cancer institutions, the American Cancer Society, and parents, but few effective treatments were available for curing childhood cancers. Conflicting themes of despair and hope pervaded *Death Be Not Proud*—the Gunthers' initial shock and fear about Johnny's diagnosis, their search for therapeutic options, their desire to preserve their son's life in the face of his shattering illness, and their faith that a medical breakthrough would save Johnny's life or those of other young sufferers.

Johnny's Diagnosis and Treatment

The opening pages of Gunther's account focused on the physicians' difficulties in properly diagnosing Johnny's brain tumor. In March 1946, during his junior year of high school, Johnny left Deerfield Academy, a boarding school in Massachusetts, and traveled to his father's apartment in New York City to enjoy a busy spring break. In the midst of attending Broadway shows and academic lectures with his parents, he received an eye exam and a "clean bill of health" from his family physician at a regular checkup.[24] Upon his return to school, a persistent neck ache coupled with the worrisome news that a classmate had been diagnosed with polio prompted Johnny to visit the school infirmary. The school physician advised rest, aspirin, and an electric pad to alleviate the pain. The relief proved only temporary, and the school physician became concerned about Johnny's symptoms. He sent him to a neurologist in nearby Springfield for a second opinion. A spinal tap revealed increased intracranial pressure, a dangerous situation requiring further tests. The physicians notified Johnny's parents, who contacted Tracy Putnam, an eminent neurosurgeon and director of neurosurgery at the Neurological Institute of Columbia-Presbyterian Hospital. Along with John and Frances Gunther, Putnam traveled to Deerfield to discuss Johnny's health. Everyone agreed that Johnny should be transported to New York City for further evaluation and, perhaps, emergency treatment. Upon his admission to the Neurological Institute, he underwent a complete set of neurological and laboratory tests. The test results confirmed that Johnny had a dangerous amount of pressure on his brain but could not supply definitive information. An accurate diagnosis could only be made after surgery and a pathological examination of the tissue.[25]

Johnny's unexpected tumor diagnosis illustrated the tensions between the prevailing messages about cancer in children and the grim reality of the disease. Experts declared that increased awareness, prompt diagnosis, and immediate treatment could reduce mortality rates from childhood cancer, but, without a distinct set of symptoms or standard diagnostic test for cancer detection, it was difficult for parents or physicians to make a correct cancer diagnosis during the disease's initial stages. Unlike other sufferers, Johnny did not initially experience the excruciating headaches, visual impairment, vomiting, imbalance, or paralysis that could signal a brain tumor. Neither the family physician, John, Frances, nor Johnny had any warning, and in their case, like many, early detection was virtually impossible. At Johnny's first operation, the tumor had already grown to the size of an orange.[26]

Gunther described Johnny's first hospitalization at the Neurological Institute in great detail. Admitted on April 26, 1946, Johnny underwent his first operation only three days later. Surgeons cut a large flap from his scalp and skull to facilitate observation and removal of the tumor. Hindered by profuse bleeding and the size and location of the tumor, the surgeons removed only half of the large mass. Macroscopic observation and pathological analysis determined that he had an astroblastoma that was undergoing glioblastomatous transformation. Such a fast-growing tumor represented a particularly grim prognosis.[27] The bone flap was completely removed to relieve dangerous compression that could build at the tumor site. Johnny had an elevated temperature during recovery, but by the seventh day after surgery the wound dressing was removed and he began his first cycle of x-ray therapy. During this time, the decompressed area bulged and physicians measured the amount of pressure on his optic nerve, known as the papilledema. His second course of radiation began on the thirty-second day after the brain operation. In the postwar era, the British focused more effort on radiotherapeutics, but American research was still at the forefront of new therapeutic device development. The cobalt bomb, cyclotron, betatron, linear accelerator, and the nuclear pile, a key source of radioisotopes, all provided new methods for delivering radiation.[28] Without a course of chemotherapy to offer Johnny, radiation was used after surgery to shrink portions of the tumor that remained and to try to reduce dangerous swelling. At the time of discharge, Johnny's medical record noted that he was "listless and drowsy, but alert mentally," had an unsteady gait, displayed a weakened muscle tone and a reduction of reflexes, but retained normal speech.[29] He completed his remaining x-ray therapy sessions as an outpatient.

During his initial hospitalization, Johnny became inquisitive about his tumor and relentlessly demanded truthful answers from his physicians, nurses, and parents at each step of the treatment process. The "truths" about the cause of his tumor, however, remained obscure. In the comprehensive text *Cancer in Childhood* (1940), Harold Dargeon theorized that physical irritants were the primary cause of childhood cancers but encouraged further research into a range of antenatal and postnatal factors, including heredity, embryonic factors, hormones, and biological mechanisms regulating cell growth. In an effort to discover clues to the rare tumor's origin, physicians questioned Johnny about any unusual blows to the head. Johnny suggested that the strain of "holding himself back" at school could have caused the abnormal growth. He also recalled a specific accident when his chair tipped backward during a chess game and he hit his head on an iron radiator.[30] The physicians acknowledged the possibilities but admitted that if they could determine an exact cause it would revolutionize medicine. On the subject of etiology, his father concluded, "Cancer causation is the greatest and most formidable of all the unknowns of modern science."[31] Thus, the Gunthers focused their attention on finding treatments, not determining etiology.

As Johnny's physical condition deteriorated, the Gunthers intensified their pursuit of possible medical options. Over the course of his illness, the Gunthers continually consulted physicians to confirm Johnny's diagnosis and to assess his condition. A physical examination by Lester A. Mount, Putnam's junior associate at the Neurological Institute, followed his second five-day cycle of radiation. Mount found that Johnny's fatigue was less marked, his strength and walking had improved, the swelling had decreased and softened, and his neurological tests gave mixed results. He concluded that Johnny had improved since he received his second course of x-ray therapy.[32] An examination on July 15, however, revealed that the tumor had resumed its growth, increased the intracranial pressure, and resulted in a thinning of Johnny's scalp.

Although a series of neurological tests demonstrated that the tumor's progression had impaired Johnny's movement, he was able to continue his normal daily activities.[33] In late July, Wilder Penfield, an eminent Canadian neurosurgeon, confirmed the diagnosis and approved of the previous surgical intervention, saying it had prolonged his life and preserved Johnny's physical and mental abilities. After conducting a physical exam and reviewing the pathology slides, he bluntly informed the Gunthers, "Your child has a malignant glioma, and it will kill him."[34] In his written report, Penfield noted that a radical amputation of the occipital lobe would not prolong the boy's life, but it might maintain his ability to function throughout his final days.[35] Facing the tumor's rapid growth and the fragile tissues at the decompression area and the lack of therapeutic options, the Gunthers suddenly felt helpless. Every physician they consulted pronounced a death sentence. Masson compared Johnny's brain to an apple with a bruised spot on it. Within a couple of months, he predicted, the organ would be consumed by the fast-growing tumor.[36] Two physicians suggested constructing a cap in Johnny's skull that would inhibit the tumor's outward growth and, rather, drive the tumor inward and speed his inevitable death.[37] Unwilling to concede defeat, the Gunthers explored other options.

John Gunther began pursuing an in-depth investigation similar to those he had conducted for his "Inside" series. Using his journalistic skills, Gunther educated himself about brain tumors and became determined to find other promising treatments or cures. At his wife's suggestion, he looked to the wartime work of physicists and atomic scientists, hoping that there were possible medical applications. He composed a standard letter that he sent to physicians and scientists at hospitals and universities across the country, including Ernest O. Lawrence, director of the radiation laboratory at University of California at Berkeley, Robley Evans, a professor at Massachusetts Institute of Technology, and Robert Maynard Hutchins, president of the University of Chicago.[38] He described Johnny's condition and the extent of his son's conventional treatment and then conveyed his unwavering determination to find other treatment options: "The war has, we all know, produced a vast intensification of research in fields like atomic fission, nuclear energy, and the like, and I am told that some specific medical developments are possible and in fact are in the experimental stage."[39] Desperately, searching for a last resort, he pleaded, "What my wife and I are wondering is whether by any miraculous chance you have come across any hint or trace of something *new* in the therapy of tumors in general, or this kind of cranial tumor in particular."[40] Gunther received replies, but they offered little hope. Spurred on by newspaper articles about intravenous mustard gas, a chemotherapeutic substance developed by military scientists, he also contacted the University of Utah, American Cancer Society offices, an experimental station in Maine, and Joseph Burchenal, a leading investigator interested in the development of chemotherapeutic agents at Memorial Hospital. Burchenal supplied experimental mustard gas for two series of treatments, but the therapy lowered Johnny's white blood cell count to dangerous levels and seemed to have little effect on his tumor.[41]

Gunther's intense search for additional treatments suggested an unwavering belief that medical science would cure his son. In *Death Be Not Proud,* he raised "a salute to all doctors" even as he acknowledged "there is much, even within the confines of a splinter-thin specialty, that they themselves do not know."[42] Calling some aspects of his son's illness complete mysteries, the Gunthers still hoped for a medical miracle for their son. Frances's words reflected this hope:

> Perhaps we tried to do too much. But Johnny loved life desperately and we loved him desperately and it was our duty to try absolutely everything and keep him alive as long as possible . . . always we thought that, if only we could maintain life somehow, some extraordinary *new* cure might be discovered. We thought of boys who died of streptococcus infections just before sulfa came into use.[43]

The Gunthers' writings revealed a therapeutic optimism, a faith in scientists or physicians to develop an agent similar to sulfa drugs or penicillin that would target cancer. Their optimistic perspective mirrored the nation's postwar enthusiasm about science, medicine, and technology. During the 1940s, the Office of Scientific Research and Development directed American scientists' studies toward wartime concerns. The success of American wartime projects like radar and the atomic bomb helped position the United States as an international economic and political leader. Penicillin production and other medical improvements related to disease prevention, treatment, and surgery had improved military health. Vannevar Bush, head of the OSRD during World War II, wrote *Science: The Endless Frontier* to summarize scientific and medical accomplishments during the war, urge postwar funding, and recommend guidelines regarding federal support of basic research at independent institutions.[44] Federal support of scientific and medical research programs continued and expanded in the postwar years as a way to address national security concerns at a time of American prosperity and optimism. Rhetoric used to describe scientific results and possible cures in newspaper and popular magazine articles assured Americans that breakthroughs were on the horizon. Investigators broadened their search for other medical wonder drugs or "magic bullets" to ensure civilian health.[45] An incurable disease, especially a disease that affected children, tempered the heroic vision of physicians' accomplishments during the "Golden Age of Medicine," but the promise of a medical miracle buoyed the Gunthers' spirits and drove their search for treatments.[46]

By the end of the summer, the Gunthers decided that "there was absolutely nothing to lose" by using alternative methods.[47] By September 1946, physicians at the Neurological Institute predicted that Johnny had only one week to live, and the Gunthers admitted their son to Max Gerson's nursing home, Oakland Manor, located in Nanuet, New York. Attacked by the American Medical Association and the American Cancer Society, Gerson offered cancer patients an alternative therapeutic regimen to surgery and radiation. He eschewed transfusions and other medical interventions and, instead, substituted rest, vitamins, frequent enemas, and the healing power of nature. Specifically, he advised a strict dietary regimen of fresh produce, no salt, no fat, and little protein.[48] Initially skeptical of the healing power of Gerson's controversial plan, Johnny's physicians grudgingly cooperated with Gerson on Johnny's case, conceding the treatment could do little harm at this stage of his disease.

The tension between the Gunthers, Gerson, and the orthodox physicians eventually intensified. In December, Gunther wrote, "So began a battle of the doctors that all but destroyed us."[49] Gerson fervently opposed a minor operation to drain an abscess on Johnny's head, but the Gunthers insisted on the procedure. As a compromise, Mount used the anesthetic of Gerson's choice and Johnny continued on the Gerson diet at the Neurological Institute. After a second dispute in which Mount insisted that he administer penicillin injections to prevent infection, Gerson wrote, "Dr. Mount's answer has to be taken as rejection of my opinion and cooperation. It is wise to fix the question of responsibility after my return, January 6."[50] The Gunthers elected to continue consulting both orthodox physicians and Gerson because Johnny's tumor had diminished in size during Gerson's care. His physicians called the cause of the regression "a riddle," but Penfield admitted, "In any case, it is encouraging and all we can do is to hang on and hope."[51] Medical reports from January to May 1947, John Gunther's notes, and correspondence between the Gunthers and physicians demonstrated that the family simultaneously relied on alternative and conventional physicians and practices for Johnny's care from the early fall through the late spring.[52] Despite their faith in the promises of postwar innovation in medicine and science, they pursued and embraced a range of therapies that might help their son.

By the end of February 1947, the tumor again began to enlarge and Johnny's condition deteriorated. He began experiencing periods of amnesia, paralysis, and partial numbness in his arms and legs. The Gunthers desperately sought a way to slow the tumor's constant, rapid growth. After another attempt to drain the abscess, Mount said, "Let Johnny do exactly what he wants to do and die happy."[53] Penfield, however, urged the Gunthers to halt the Gerson diet and attempt another cycle of therapies—a surgical intervention, x-ray therapy, nitrogen mustard injections, and a diet regimen. After the operation, Mount told the Gunthers that he extracted two handfuls of material from the tumor site but that he had not reached healthy tissue at a depth of eleven centimeters.[54] John later admitted, "It was simply impossible to let this child die" and recounted the additional phone calls made and letters sent after receiving this pessimistic news.[55]

As his parents continued their search for a life-saving treatment, Johnny became increasingly driven to complete his schoolwork, pass his final examinations, take the college board exams, apply to Harvard, and graduate with his class at Deerfield in June. He spent time at the public library and attended the New York Tutoring School. He traveled to Deerfield to accept his high school diploma and participate in graduation weekend events. On graduation day, he walked up the aisle and personally received his diploma to thunderous applause from his classmates and their parents. Less than a month later, on June 30, 1947, Johnny Gunther suffered a cerebral hemorrhage and died at 11:02 in the evening.[56]

Reading *Death Be Not Proud*

John Gunther noted that he and Frances received an "avalanche" of telegrams and letters following his son's death and the wide distribution of his obituary by the Associated Press.[57] He wrote,

> There were condolences from camp counselors who had not seen him for years and from a barber in a downtown hotel; from the Negro elevator boy in my office building and the proprietors of a friendly restaurant on Madison Avenue; from playwrights, judges, politicians, old Chicago friends whom we had not seen in years, from teachers and doctors and newspaper folk, old schoolmates, several of those who had seen him graduate at Deerfield, movie people, poets, acquaintances from far-off days in Vienna, physicists, his godfather in Washington, the doormen at our apartment building, refugees from Europe, his devoted governess Milla, scores of writers, and above all people who had never met him or us—parents whose sons had also died.[58]

They also received thousands of letters after the publication of *Death Be Not Proud* as a condensed article in *Ladies' Home Journal* and *Reader's Digest* and its release as a full-length book.[59] The number of letters and telegrams of condolence that friends and colleagues sent to the Gunthers attested to the family's personal and professional stature, but the hundreds of responses from anonymous male and female readers of all ages and geographic regions suggested that *Death Be Not Proud* held universal appeal.[60]

Why did the book attract such a wide audience and why did so many readers feel compelled to respond? Gunther's reputation as a best-selling author contributed to the book's popularity. Favorable reviews printed in popular magazines and local newspapers may have attracted others. Reviewers praised the volume in descriptive terms—heroism, strength, and courage—redolent of the recent war. They compared Johnny's last months to a series of valiant battles in which Johnny was victorious.[61] Another group of readers may have heard about the book from friends, family, or coworkers. Readers wrote passionate letters saying that they had completed the entire article or book in one sitting, reread the story several times in different versions, and felt an irrepressible urge to write to the Gunthers. Americans also encountered the book during religious services, at the doctor's office, or in the classroom as ministers, physicians, and educators read and reacted to *Death Be Not Proud.*

Religious leaders drew on the book for life lessons. In April 1949, the minister at Community Church of New York delivered a sermon titled "Life's Eternal Challenge: A Sermon on John Gunther's 'Death Be Not Proud.'" As he explained to Gunther, Nathaniel M. Guptill, minister at First Congregational Church in South Portland, Maine, delivered a sermon based on human suffering as a key to salvation.[62] A Unitarian minister in Indianapolis, Indiana, speaking on a similar theme included *Death Be Not Proud* in his Easter radio address.[63] Minister Niels Nelson wrote to the editor of the *Ladies' Home Journal* explaining that he had purchased twenty-five copies of the article to distribute to his parishioners and planned to buy more. He referred to the magazine's decision to print the article as a "fine public service" and equated the piece with John Hersey's "Hiroshima" printed in the *New Yorker.* "Because of this account alone Johnny has not lived, or died, in vain," he wrote, "Because of him others will strive toward valiancy."[64] Ministers saw Johnny's life as a model for their congregations.

Physicians, medical students, and nursing students wrote to the Gunthers about the medical portions of *Death Be Not Proud.* Paul B. Hoeber of the medical book department of Harper and Brothers characterized physicians as a "skeptical lot in their approach to lay writing on medical subjects [who] must harden themselves to the human emotions to which illness and death must give rise" but reported that they had overwhelmingly approved of the book.[65] Doctors praised the accuracy of the medical sections and acknowledged the excellent care that Johnny received. Walter L. Palmer, professor of medicine at the University of Chicago, credited both the medical staff and the family's devoted care for his extended survival, saying, "Johnny received the best medical care possible."[66] Palmer also approved the Gunthers' approach toward Gerson's alternative diet regimen. He noted that "from a strictly scientific point of view the diet was of no value" but did not criticize the Gunthers for employing this unorthodox method as a last resort.[67] Perhaps predictably, the Gunther's use of alternative therapy did divide medical readers. Charles Huggins, a researcher in the Ben May Laboratory for Cancer Research at the University of Chicago and later a Nobel laureate for his work on prostate cancer, vehemently disagreed with Palmer's assessment, urging Gunther to "re-issue your classic omitting all Gersonia."[68] He noted that the American Cancer Society could use a revised edition to promote cancer research. Correspondence between the University of North Carolina and Gunther, however, showed that an audio version of *Death Be Not Proud* was broadcast on behalf of the regional American Cancer Society chapter in 1949. The information on Gerson remained. Most physicians who wrote letters commended Johnny's care and did not note particular disagreement with the Gunthers' use of alternative therapies.

Letters from medical and nursing students suggest that *Death Be Not Proud* challenged them to question their training and provided guidance on how to approach similar cases in the wards and in the classroom. One fourth-year medical student at Yale University asked whether the Gunthers would have reconsidered their decisions if they had found Johnny's "basic human qualities slipping away."[69] She hypothesized that aggressive treatment may not be the best course for all patients with brain tumors. Recognizing that this decision-making process would be a formidable challenge during her career, she asked hypothetically, "How can one eliminate the statistics and treat the particular?"[70] Mary L. Davis, a graduate nursing student in California who had accepted a position as a nursing instructor after graduation, hoped to teach her students about their role in times of patient and parental grief. Davis thought it would assist family members if the grieving process began before a child's death. She asked Gunther how nurses could have been more beneficial to him, his wife, or his son during Johnny's illness.[71]

Educators who wrote to Gunther viewed *Death Be Not Proud* as an invaluable resource for teaching their students about their character and their relationships with others. Jean Kuffner, a high school teacher from Greensburg, Pennsylvania, wrote that she liked the way Gunther revealed Johnny's "personality, selflessness, and devotion to his parents."[72] She reported that the *Reader's Digest* excerpt held the attention of each member of her high school class. Lois Payne, an English teacher from Sheridan, Oregon, expressed that she wanted *Death Be Not Proud* to be on the shelves of every high school library across the country.[73] Mrs. D. B. Hopkins, a resident of Bath, Maine, who had lost her seventeen-year-old daughter Betty to a brain tumor, wrote, "My sister is a book buyer in a large book store in Portland, Maine. She said that the demand for your book is tremendous—that college students especially were coming in to buy it."[74]

Dozens of students from across the country wrote to the Gunthers after reading *Death Be Not Proud* for school assignments or during their free time. Some admired Johnny's determined fight to continue his education throughout his illness, some gleaned life lessons from the story, and others simply chose the book because it featured a boy their age. Betty Jane Sheffler, a student from Greensburg, Pennsylvania, read the article in her English class. She wrote, "I especially liked the way Johnnie [*sic*] thought of his parents and not himself when Dr. Putnam told Johnny about the operation for a brain tumor."[75] Mary Reilly, a teenager from Toronto, claimed it was the best book she had ever read: "I could almost, well I could see him, feel him, almost as though I had just been talking to him, and then to you and Frances."[76] She and others said that they had encouraged all of their friends to read it, too. In 1958, teenager Beverly Goodell from Stafford Springs, Connecticut, wrote that *Death Be Not Proud* had challenged her to look beyond her own concerns: "If parents read your book," Goodell explained, "they would get a better viewpoint in which they would realize how teenagers feel toward their parents. Perhaps there would be better understanding between parent and teenager."[77] Only a few students who wrote could relate personally to the challenges presented by a life-threatening illness, but they shared Johnny's concerns about schoolwork, parents, and uncertainty about the future. Young people identified with Johnny as they might relate to a peer.

Parents as Letter Writers

Parents, the largest group of letter writers, empathized with the Gunthers' experiences and shared their personal experiences with a child's illness, disability, or death. Like John and Frances's sections, readers' responses to *Death Be Not Proud* exemplified the accepted gendered division of parental behavior in the late 1940s. By combining two complementary, gender-specific perspectives, *Death Be Not Proud* appealed to both men and women. One mother wrote, "Neither one of you alone could have accomplished what you have done together. Its message could not have been delivered by one sex without being balanced by the other—like so many things in life."[78] Another related, "I was as brave as John senior during his account, but when I read through a mother's heart and soul the tears flowed and won't stop even now."[79] Although many fathers wrote Gunther to inquire about medical treatments and to ask his opinion on the efficacy of the Gerson diet, many more mothers wrote after reading *Death Be Not Proud.* Hundreds of mothers responded to the story's first appearance in the *Ladies' Home Journal* in which John's name was in the byline.

Letters designated specifically for Frances were almost exclusively from mothers. Mothers' letters resonated with emotions expressed by Frances. The same sentiments—those of guilt, intimacy, and maternal love—that John Gunther labeled as too personal moved hundreds of women to write. Mothers addressed "A Word from Frances" directly. Many readers cited Frances's statement, "I wish we had loved Johnny more."[80] They assured Frances that her child-rearing methods had served as an inspiration for their daily interactions with their own children. Jane Raborg wrote, "I doubt that you will ever know the far-reaching effect upon child-parent relationships which your book, and especially the 'Word from Frances' shall have."[81] Raborg said that she felt newly grateful for her own blue-eyed, blond-haired boy. Mothers also empathized with the uncertainty, frustration, and guilt Frances expressed in her section. Marianne Peters, a mother from Chicago, despaired that her daughter had just been diagnosed with diabetes. She questioned, "How is it possible? She couldn't have had better care! She just had a complete check-up last month!"[82] Peters illustrated that mothers felt a measure of personal responsibility when their child suffered from serious disease or died. Through their letters, parents assured John and Frances and, conversely, themselves, that they had done everything possible to nurture, treat, or cure their own child.

Overwhelmingly, parents wrote to share their own experiences of illness and loss with the Gunthers. They frequently attached copies of obituary notices, funeral service bulletins, poems, and photographs of their dead children to their letters. Some made comparisons between Johnny's appearance—illustrated in a photograph printed opposite the title page—and their own children's features. For most readers, correspondence with the Gunthers ended after sending and receiving one letter. John Gunther replied to most letters with a short note that simply expressed his gratitude. Other parents asked for treatment advice or included specific requests for autographed copies of article or book, donations for cancer research programs, financial help to cover medical bills, and tips on how to write and publish their own experiences with illness. In select cases, Gunther tailored his replies to individual requests. In a few instances, though, the original writer sent additional correspondence. In one extended exchange, Elizabeth V. Guthrie, a widowed mother from Brigantine, New Jersey, wrote a series of letters about her daughter Patty after reading about Johnny's illness in the *Ladies' Home Journal.*

Guthrie explained that six years earlier, in May 1943, her fifteen-year-old daughter Patricia (Patty) Guthrie underwent a brain tumor operation at the University Hospital in Philadelphia after physicians detected diminished eyesight and abnormalities in her pituitary gland function. After the surgery, Guthrie carefully selected her daughter's food, encouraged her to rest and enjoy the sunshine, and restricted strenuous exercise such as "riding horseback, tennis or diving or running fast" with the hope that the tumor would not recur.[83] Every six months Patty had a regular appointment for a checkup and to test her vision. In 1949, a hemorrhage blinded her right eye and indicated that a tumor was present. On November 18, 1949, Patty had a second operation, but the tumor could not be completely removed due to its position. Physicians predicted that she had two months to live.

Guthrie's other children cautioned their mother not to buy a copy of *Death Be Not Proud,* but she told the Gunthers that she found it comforting to mark every aspect of Johnny's illness that mirrored Patty's experiences. She also felt "so much better after pouring out my soul to someone who understands."[84] John Gunther responded by letter within a month. Guthrie promptly thanked him for his reply and related that Patty was "thrilled" to hear of the response. Approximately a year and a half later, on June 25, 1951, Guthrie sent a long letter updating the Gunthers on her daughter's declining condition. She wrote that her daughter had vomited nearly very day for a year, suffered from convulsions, memory loss, chills, and fatigue, and relied on morphine to provide relief from her symptoms at night. She reported that Patty also had periods of wellness when she felt strong enough to sit on a deck at their beachfront home, which she had bought for the comfort of her dying daughter. Quoting an earlier letter from Gunther in which he said he "hoped the future would be kind" to the Guthries, she replied, "It has, in a way, as I have Patty and will for a month, or so."[85] Guthrie soon reported that Patty's health had deteriorated further and that they had hired a trained nurse and acquired a hospital bed and a contoured chair to aid in her daughter's care. She repeated, "I want you to know how much your book has helped me to carry on" and noted, "Somehow I get comfort in talking to you."[86] Three months later, on October 25, 1951, less than two weeks after her daughter's death, she informed the Gunthers that Patty had died.

Guthrie's correspondence differed from other letter writers in that she sent the Gunthers a series of letters that kept them abreast of her daughter's condition. Guthrie tried to create a dialogue, a two-way conversation with other parents who could empathize with her struggles and her daughter's suffering. Although the number of letters she sent was exceptional, she represented the majority of the parents in that she related closely to the events in *Death Be Not Proud* and looked to the book for guidance in medical matters. In an era before support groups for grieving parents, Guthrie and other letter writers sought the solace of fellow sufferers through their letters.

Etiology

Surprisingly few parents wrote to the Gunthers with questions or theories about cancer etiology. In one exception, Mrs. Martin H. Byrd from Wadesville, Indiana, conveyed uncertainty about the cause of her sixteen-year-old son Kenny's brain tumor. She offered physical trauma as a possible explanation, as had been done in Johnny's case. She wrote,

> Whatever caused this to happen we'll never know although he had several small accidents as children do, falls as a small child and was hit by a ball while playing baseball and fell at High School when a freshman and hit his head, or where or when its just hard to believe anything could cause so much damage. Even perhaps when he was borned, an unusual delivery, may be the answer. We'll never know.[87]

After reading an excerpt of the Gunther's story in *Reader's Digest,* Mrs. T. H. R. sent a letter reflecting on her younger brother Albert's death from a brain tumor thirty years earlier. After receiving the diagnosis, their father had taken Albert from their rural home outside of El Paso, Texas, to Boston for a consultation with Harvey Cushing, the leading neurosurgeon of the time. Cushing deemed the tumor inoperable and the boy soon died, but the family remained curious about the cause of the mysterious growth. Albert's physicians had suggested that a severe head trauma could have caused the tumor, but family members could not pinpoint an exact incident. However, Mrs. T. H. R. wrote, "In my brother's case it could have been a kick by a burro which it was a habit of his to ride. But we will never know exactly what."[88] Explanations that implicated physical irritation or trauma as a major cause of their child's disease may have appealed to parents because they could pinpoint particular incidents that could have caused tumors, making a mysterious, seemingly spontaneous illness more understandable.

For most writers it was the varied causes proposed by investigators, yet a paucity of definitive "expert" knowledge that was the most upsetting.[89] Losing her eleven-year-old daughter after only a few months of cancer treatment, Mrs. M. J. S. asked, "Why can't medical science start learning how these things develop, where they come from? It seems they just appear very suddenly out of nowhere."[90] These letters and others demonstrated that during the 1940s and 1950s there was a wide gap between parents' need for information and the scientific evidence available about researcher's prevailing theories of cancer causation.

Diagnosis and Treatment

Most parents who wrote about their children's illnesses touched on two dominant themes in John Gunther's section of *Death Be Not Proud:* diagnosis and treatment. Many described the difficulties they experienced obtaining an accurate diagnosis for their child's cancer. Mrs. Lewis Orrell, Jr., of Klamath Falls, Oregon, expressed frustration with a year-long series of delays that hampered the diagnosis of her fifteen-year-old son's brain tumor.[91] In December 1948, Orrell took her son, George, for a physical examination. The physician recommended that they remove George's tonsils during Christmas vacation. In February, she noticed that her son had an unusual pallor. At this visit, the physician diagnosed a thyroid condition and prescribed medication for him to take for the remainder of the spring. In early fall, George displayed a strange movement in his right foot. Again, he received a prescription, but strange symptoms continued: George's handwriting faltered, he complained of ringing in his ears, and he felt faint when moving from a sitting to standing position. When he returned to the doctor, the new diagnosis was progressive polio. In the week before Christmas, George suffered from an attack of severe vomiting and headaches. The same physician recommended sulfa drugs to cure a sinus infection. Two days later (and a year since the initial exam) George's mother returned to the doctor's office with her son and asked if he might have a brain tumor. She described the exchange in her letter:

> He laughed at me and said to get it right out of my mind—it was the silliest thing he ever heard. I got mad then and told him I thought he was guessing. He said, "Now don't get mad, I think you are over-emotional about this. I think we should have the Child Guidance clinic look at him."[92]

At this point, Orrell sought another opinion. The second physician found a large tumor and immediately scheduled an operation. Surgeons were not able to remove the entire mass. Upon the physician's recommendation, Orrell read the condensed version of *Death Be Not Proud* in the *Ladies' Home Journal* to help her prepare for the challenges that her son's brain tumor could present. Orrell's experience illustrated that mothers sought to protect their children but were impeded by cancers' vague signs and symptoms and physicians' paternalistic attitudes toward their patients.

Readers from such disparate locations as New Zealand, Europe, India, South America, Canada, and across the United States from New York to Hawaii were united by common bonds—parenthood and childhood illness. Most shared a hope in the promise of scientific and medical research to treat and cure their child. Letter writers such as Mrs. Cramer from Woodstock, New York, drew close parallels between Johnny's treatment and her own child's experiences.[93] Mrs. Cramer's twenty-year-old son, Chico, also endured multiple hospital stays, conferences with experts in the field, an attempt to follow the Gerson method, "mustard" doses, and brief respites at home. In another case, Mrs. Raymond Kaplan wrote, "Just three weeks ago, with no warning whatsoever, my little son was stricken with infantile paralysis and died within 48 hours."[94] Like John and Frances, she wished for a medical breakthrough. Kaplan wrote that through science "these two horrors, polio and cancer will be under control."[95] Like the Gunthers, readers combed newspapers and magazine headlines for breakthroughs and surveyed scientists and physicians for promising early developments.

Unlike the physicians who disagreed about the Gunthers use of alternative therapies, parents admired the Gunthers' ability to find innovative conventional and alternative cancer therapies and pleaded with them to share their findings. Much like John Gunther's letters to experts during Johnny's illness, parents filled their letters with a detailed medical history, provided a summary of the treatments they had already exhausted, and described their desperation. They clearly respected the Gunthers' knowledge of the field and considered them to be lay experts. J. Davis, whose daughter Iris had also been diagnosed with a brain tumor, wanted Gunther's advice. After surgery and radiation, his daughter had slipped into a coma and her left side was paralyzed. Physicians had sent her home to die. Davis—certain that additional treatments existed—told Gunther that he, too, had sent out a worldwide appeal to find a cure. He discovered promising results from a radioactive phosphate research program at the University of Wisconsin and had pressured the British Medical Research Council to contact the U.S. Atomic Energy Commission for a supply. After the isotope injections, her physicians reported that her tumor had been destroyed and part of her brain had suffered permanent damage. After reading *Death Be Not Proud,* Davis eagerly enlisted Gunther in his ongoing search to find advice on isotopes and other promising therapies.[96] Readers seeking alternative treatments also pleaded with Gunther to reveal his opinion of Gerson's role in Johnny's temporary improvement and asked him to send Gerson's address so that they could contact him and present their child's case.[97] Readers viewed John Gunther and *Death Be Not Proud* as authoritative, yet accessible resources for learning about medical treatments.

The majority of parents related closely to *Death Be Not Proud,* but some highlighted differences between the Gunthers' circumstances and their own. A few letters underscored financial disparities. One mother wrote, "When people of a limited income have this kind of illness and although a very reliable brain surgeon did the operation, one always feels if money could help, but your experience and your writing has helped so many to know that not too much can be done."[98] Despite her economic constraints, Roessler found comfort in *Death Be Not Proud.* It placated her fears that additional, expensive medical procedures would not have saved her child. Parents such as Genevieve Christiansen from Boise, Idaho, expressed reservations about enlisting medical science or exhausting all options in a child's medical care, even if it was available and affordable. As the family of a best-selling author residing in New York City, Johnny and his family had privileged access to medical experts and services. They relied on friends and colleagues to give second opinions on Johnny's case, suggest experimental therapies or promising research, and help them publish their book. Christiansen accused the Gunthers of unnecessarily prolonging Johnny's life:

> [You] throw yourself on experts, aha they will solve your problems but with all their reputations of knowing all, they too are human and know very little more of the nature and intricacies of human diseases and cure than you do. Even tho [*sic*] they encourage their practice with a surety and confidence. It all seems a hodge podge of trial and error without consulting the person involved whether he wants all the extra pain inflicted on him. I think we parents get a little bit crazy at such a time.[99]

Christianson regretted that she had submitted her daughter to the medical procedures and hospitalization the girl endured during her last days. Her letter disclosed that her daughter had died of a mastoid infection (related to an inner ear infection) fourteen years earlier and felt that her trust in the advice of medical experts had been betrayed. Her letter expressed the bitterness she felt after the loss of her own child. Christianson concluded her letter, "I sometimes wish folks like us could organize and get together, for there are still so few of us. It would take away the loneliness."[100] Despite their different experiences, Roessler, Christiansen, and others joined other parents in writing letters to the Gunthers. *Death Be Not Proud* created an outlet for all parents to communicate their family's experience of cancer, illness, death, and grief with other empathetic parents at a time when there was little public discussion of such subjects.

Child Loss and Parental Grief

Readers repeatedly declared that nothing was more devastating than losing a child. Why did the death of a child cause parents such profound anguish in the mid-twentieth century? The sentimental value of children, a "reaffirmation of domesticity" and the promotion of distinct roles for women and men that followed World War II set the stage for intense parental grief.[101] As women were urged to relinquish their jobs to returning servicemen and focus their attention on motherhood and maintaining a home, "housewife" and "mother" were lauded as rewarding, natural occupations for women. Marriage, family, and the home were praised as secure havens in a prosperous but uncertain time in the United States. A dramatic "baby boom" contributed to a new culture that formed around motherhood, child rearing, and a growing number of consumer goods promoted as essential for the domestic sphere—a car, television, refrigerator, and washing machine.[102] Children became central to family life and symbolized the promise of the future. Wendy Kozol has shown that images published in *Life* in the 1940s and 1950s consistently promoted the white, middle-class, nuclear family as "quintessentially American" and representative of a national ideal.[103] The death of a child threatened the structure and stability of this postwar family.

Death Be Not Proud encouraged readers, especially mothers, to reflect upon and express their private grief over losing a child. After the loss of her twelve-year-old son six months earlier, Hieda N. Janovak had been "unable to shed a tear, nor, hardly to discuss it" and she wished that her own life would end.[104] After reading *Death Be Not Proud,* she claimed, she could not help but write a letter. William Rodgers of Ossining, New York, recognized the value of *Death Be Not Proud*'s story of illness and death. He told Gunther that friends, a young couple, had recently buried their son: "I shall give the parents of that little boy my magazine to read. I think it will comfort them, but more than that I think it will enlarge and enrich their view—as it does mine—of all living and dying."[105] In the mid-twentieth century, few publications addressed child loss or provided an outlet for parental grief.[106] *Death Be Not Proud* helped fill the void in literature about child death and parental grief in the 1940s and 1950s and served as sources of community and comfort in a time of despair. The book and the responses that followed created connections between parents who felt isolated during a child's illness and after his or her death.

Johnny's obituary and the story of his death rekindled parents' memories of young soldiers who lost their lives during the Second World War. Mothers who had mourned the death of their sons identified with the Gunthers' loss and suggested that losing a son, an only son, or an only child caused parents additional despair. Like the book's reviewers, they used war rhetoric to draw parallels between a valiant wartime death and Johnny's courageous death from cancer. Mrs. A. J. Hummel from Utica, New York, wrote, "A lot of people, that is parents, lost an only son in the war and I suppose you'll say that you wouldn't mind so much if *he* had died for his country. I think Johnny died as brave as any soldier on a battlefield and perhaps more so."[107] Loretta Maxwell lamented that she was not able to be at her son's side "as he suffered and died on a battlefield only a few years older than Johnny."[108] Helen L. Kaufmann shared that her son was killed in World War II after three years of imprisonment in a Japanese camp. Unlike the Gunthers, "There were no last months of companionship to create a bulwark against loss. There is no grave to visit or weep over."[109] She reflected that, perhaps, the Gunthers' ordeal ultimately had its compensations. Margaret J. Oberfelder of New Rochelle, New York, wrote that she and her husband had lost two sons—one with a mastoid infection and the second in a B-17 crash during the war. She wrote, "The natural order of life is for we parents to live on in our children and it takes great courage for us to keep them alive in us after death."[110] In his fight against cancer, Johnny had acted as a courageous soldier and a hero on a metaphoric battlefield that had caused similar hardships to those that the war had caused their sons, their families, and America.[111] Like war memorials, *Death Be Not Proud* honored a victory over a formidable, physical enemy, invited public mourning, and confirmed the immortality of the dead.

Parents' responses to *Death Be Not Proud* showed that John's and Frances's sections affected parents deeply. On a fundamental level, parents realized that there was a larger community of child sufferers and bereaved parents. Mrs. G. Clifford of Toronto, Ontario, wrote, "I thought at the time our child must be the only child so afflicted."[112] Parents like Lucy P. Gregg of Hastings-on-Hudson, New York, were particularly compelled by the book's discussion of death. She admitted that death was a topic "so few of us are willing to face."[113] Expressions of child loss and parental grief found an outlet through the pages of *Death Be Not Proud* and readers' letters. Gregg also admired the Gunthers' ability to immortalize Johnny and expose the "potentialities in family life" through their dedication to their son.[114] While *Death Be Not Proud* publicly addressed the challenges of terminal illness and death, readers' responses revealed that they viewed such events as a threat not only to the child but also to the postwar ideal of the nuclear family.

The Gunthers, who had divorced years earlier and only had reunited temporarily to manage Johnny's illness, were a precarious "family" according to postwar definitions. Readers enthusiastically urged them to restore their marriage and to devote their attention toward another child to cope with their loss. By replicating the ideal postwar family, readers insinuated, the Gunthers could find relief and contentment. Frances Gordon enclosed a photo of her son, Chuck, and included several letters that he had sent home during his first year of boarding school. Upon his mother's request, Chuck wrote his own letter to the Gunthers, suggesting that they act as honorary "foster parents." He wrote, "If you'll let me, I'll write you both once a week for awhile in the same vein that I would write my parents (as I do)."[115] Chuck sent another rambling letter less than two weeks later, but the correspondence then halted abruptly.

Gertrude Hepworth from Larchmont, New York, suggested that she lend her son Malcolm to the Gunthers for a Saturday or Sunday. She closed her letter with the hope that "many parents will be helped in accepting sickness and sorrow" and, like her, "many might learn the real value of family relationship."[116] Other parents who had lost children suggested adoption. Mildred Mize from Bellingham, Washington, urged the Gunthers to support Bill Gunter, a youth from a poor family whose academic interests were in speech and science.[117] Mize included a news clipping from the *Seattle Post-Intelligencer* depicting Gunter and his teammates explaining atom smashing in their prize-winning speech at the state oratorical competition. Readers hoped that by adding another son the Gunthers' grief would lessen. The introduction of another child ensured renewed happiness and security after an incurable or chronic illnesses such as cancer threatened the family structure.

Death Be Not Proud became a literary classic.[118] Letters from parents slowed, but correspondence sent by students revealed that teachers continued to assign the book in their classrooms. How *Death Be Not Proud* attracted and maintained such lasting attention is a story in itself. The book reached a wide audience through John Gunther's publishing fame and the Gunthers' frank portrayal of their family's circumstances. Although the Gunthers could not accurately be described as representative of an "average" American family in the 1940s, through Johnny's own words and his parents' careful descriptions of their son, he became "every boy" to many readers. A radio script based on *Death Be Not Proud* began,

> You've known this boy. You've seen him many times . . . you can see him now. This boy? . . . No. Some other boy perhaps. Some other boy who is learning to ride a bicycle . . . who clutters up the house with treasured "contraptions" . . . some other boy who watches beside a waterfall or catches bullfrogs in a camp . . . some other boy, not this one. For this was Johnny Gunther . . . and he is dead.[119]

Readers of all ages closely identified with Johnny and viewed him as a model American, student, teenager, son, and patient. Many letter writers proclaimed that they "loved" Johnny. In a recent essay, Viner and Golden wrote, "We must ask how children's experience of care and treatment and their lives and deaths have changed the practice of medicine" and, more broadly, how their experiences shaped culture.[120] *Death Be Not Proud* posthumously authorized Johnny Gunther's narrative and publicized his story. Like polio poster children and young cancer patients who wielded the American Cancer Society's symbolic Sword of Hope at annual meetings, Johnny Gunther highlighted cancer's threat to people of all ages. As shown through the examples of Robert DeVilliers, Jimmy, and John Gunther, the stories and voices of individual children were powerful agents to alert and educate the public about cancer.

Johnny's prolonged illness and the Gunthers' ultimately futile search for effective treatments dramatized the need for increased cancer research in the 1940s. John Gunther's narrative resembled a medical record in its chronologically organized, detailed description of Johnny's care. Gunther powerfully illustrated the postwar tension between the hope for a medical breakthrough and the fear that cancer was incurable. A rare and incurable brain cancer threatened to undermine their belief in scientific and medical progress. They nevertheless continued to seek conventional medical treatments and experimental therapies during each stage of Johnny's illness. By writing that they planned to donate the book's profits to cancer investigation, the Gunthers publicly demonstrated their unwavering faith in postwar research to find a cancer cure. Letters from parents mirrored the Gunthers' story. Faced with an ill or dying child, parents exhausted the available medical options. Despite the limits of medicine, parents embraced its promise. *Death Be Not Proud* joined the Leukemia Society of America and the Jimmy Fund in advocating childhood cancer research at a time when national funding for research was increasing dramatically.

"A Word from Frances" resonated with many mothers during the postwar baby boom. Like Benjamin Spock's *Common Sense Book of Baby and Child Care,* Frances's words alleviated mothers' concerns about proper child-rearing practices. Their letters revealed that they admired Frances's methods and planned to replicate them in their own families. Readers struggling with a child's illness or death found solace in Frances's experiences, unresolved questions, and public mourning. In response, parents recounted their own stories of despair and loss. Through their correspondence, they created a dialogue between mourners at a time when there were few forums for expressing parental grief.

A rich narrative told from three perspectives, *Death Be Not Proud* documented the Gunthers' attempts to overcome cancer's menace through medical means and to deal with their child's death. The book also had a profound impact on parents who first recognized cancer and incurable childhood illnesses as a threat to the ideal child-centered family, a safe haven during an uncertain postwar period. *Death Be Not Proud* and its careful recounting of the Gunthers' experiences continued to be relevant for parents facing similar challenges and decisions in the years that followed. In an extended letter to John Gunther, Angela Burns described her daughter Mary Sheila's diagnosis and her short, but intensive treatment by Sidney Farber in Boston. Her letter suggests a range of familial concerns—only visible through personal sources—that accompanied the discovery and administration of a number of new, experimental therapies for acute leukemia in young patients.

CHAPTER FOUR

"Against All Odds"

CHEMOTHERAPY AND THE MEDICAL MANAGEMENT OF ACUTE LEUKEMIA IN THE 1950S

Only a few weeks after her six-year-old daughter, Mary Sheila, died of acute leukemia, Angela Burns and her husband read *Death Be Not Proud.* Moved by the Gunthers' story, the grieving mother sent John Gunther a seventeen-page letter in which she recounted her daughter's diagnosis, her nine months of treatment at the Jimmy Fund Clinic in Boston, and her death in January 1949.[1] Like the Gunthers, Burns lamented that, although they had tried "against all odds to save our child," the diagnosis of acute leukemia was a death sentence for Mary Sheila and for all children diagnosed like her in the 1940s and 1950s.[2]

Cancer remained a frustrating medical enigma at a time when science and medicine were publicly lauded for producing "magic bullets" to cure a comprehensive list of acute, infectious diseases, including tuberculosis and diphtheria.[3] The development of polio vaccines reduced parental anxiety caused by this highly visible, potentially disabling disease. In addition, the introduction of broad-spectrum antibiotics provided dramatic evidence of medical achievement by reducing the public dangers of sexually transmitted diseases and altering morbidity and mortality rates from childhood killers. During World War II, a collaborative research group composed of members from government agencies, universities, and pharmaceutical companies from the United States and Britain had conducted pioneering research in microbiology, and, as industry recognized the potential profits from this new field of prescription medicine, they broadcast the news of these revolutionary drugs.

Although improved nutrition, housing, and living conditions may have also contributed significantly to the declining rates, sulfa drugs and penicillin helped bring diseases that were once considered "incurable" under control in industrial countries. During this decade and the next, attention shifted from acute disease to chronic conditions.

For decades, however, cancer resisted a chemotherapeutic breakthrough because it was generally perceived as a localized disease treatable through surgery or irradiation, not a systemic illness that would respond to drug therapy. Those involved in this area pursued two lines of research: one group tried to elucidate the mechanisms of basic cell growth, while the other identified and tested possible chemotherapeutic agents against tumors in animals. The first method was viewed as challenging, but more rational and practical than the far-reaching search for effective drugs. With the discovery of the structure of DNA in 1953 by James Watson and Francis Crick and the development of novel theories about cell division and replication, some investigators deemed this path more promising. They believed this knowledge of cell structure and function would enable them to separate the "normal" from the "pathological" and then develop appropriate therapeutics. The second group adopted an approach that was criticized as too random but had the potential to yield major rewards. Despite this discord among scientists, popular publications encouraged the public to remain optimistic that with adequate funds, staff, and facilities a "magic bullet" to cure the disease would surely be discovered.

During the late 1940s and 1950s, research on acute leukemia yielded several unanticipated but promising chemotherapeutic agents—drugs that held the potential to control the growth of or destroy cancer cells throughout the body. In response, physicians and scientists mounted a collaborative, national research program to screen and develop additional chemical therapies.[4] This chapter traces the scientific development of antileukemic agents and considers how the advent and proliferation of cancer chemotherapies profoundly shaped the medical management of the disease. Personal correspondence, memoirs, popular newspaper and magazine articles, and American Cancer Society publications illuminated the evolving relationships between biological disease, pharmaceutical development, and the changing experiences and expectations of young patients with acute leukemia and their families in the postwar period.[5]

Warfare, Chemotherapy, and Acute Leukemia

The discovery and development of mustard gases—chemical warfare compounds that later proved to be effective against cancer—began during World War I.[6] Although many countries had signed the 1925 Geneva Protocol that pledged not to deploy chemical weapons, American military intelligence maintained throughout World War II that chemical warfare was a potential threat. Consequently, investigators directed their work toward defensive measures such as creating novel methods to administer existing blistering gases, fabricating new protective masks and clothing, and formulating and testing new chemical agents that could be deployed safely by American troops.[7] Chemical research on trench warfare weaponry returned to World War I studies on mustard gas poisoning that had reported that exposure to this family of gases caused delayed, negative changes in blood and bone marrow.[8] During World War II, under a contract between the Office of Scientific Research and Development and the Department of Pharmacology at Yale University, researchers studied the pharmacological effects of nitrogen mustards, compounds closely related to the substances previously employed.

Wartime research on nitrogen mustard compounds contributed to palliative measures and treatment methods for leukemia, lymphoma, and other diseases of the blood and blood-forming organs. At Yale, Louis S. Goodman, Alfred Gilman, Frederick Philips, and Roberta Allen investigated the cytotoxic properties of the gas and determined that its effects resembled the cellular effects of x-rays on a number of tissues. Encouraged by data from experiments in which nitrogen mustard caused temporary regressions in mouse tumors, the group pursued clinical trials of nitrogen mustard as a cancer chemotherapy agent in humans. In December 1942, nitrogen mustard was injected into a patient suffering with an advanced lymphosarcoma at New Haven Hospital. After his tumors regressed, the compound was administered to additional patients. After observing favorable results in the preliminary studies, clinical trials with nitrogen mustard were begun with cancer patients of varied ages at Memorial Hospital in New York City, the Billings Hospital of the University of Chicago, and the medical school at the University of Utah in Salt Lake City. By 1948, nearly 150 patients with lymphosarcoma, Hodgkin's disease, leukemia, and other related conditions had been treated.

In 1946, Goodman and his colleagues published a report in the *Journal of the American Medical Association* that described the preliminary results of the intravenous nitrogen mustard studies.[9] Many of the patients had already undergone radiation therapy, were resistant to further roentgen irradiation, and were in an advanced or terminal stage of their illness. Despite their poor prognosis, researchers observed dramatic remissions in patients with Hodgkin's disease—partial or complete disappearance of large, numerous tumor masses. Nitrogen mustard also relieved patients' symptoms, reduced fevers, and restored their appetite, weight, and strength, allowing them to temporarily return to work for weeks or months. The investigators called for further research into the optimal dosage, dosage schedule, and combination of mustard agents.[10] C. P. Rhoads, the chairman of the National Research Council's Committee on Growth, issued an official statement in the same prominent journal advocating the use of nitrogen mustards as part of a cancer treatment program.[11] He emphasized that the tumor regressions produced by the nitrogen mustards were only temporary and did not lead to a cure but proposed further laboratory studies on related compounds and on cancers of different structure or physiology. Nitrogen mustards acted as the first step in the development of myleran and other chemotherapy agents that later proved to be valuable in treating children with acute leukemia.

Sidney Farber, Aminopterin, and Mary Sheila

Soon after the nitrogen mustard results were published, Sidney Farber announced his findings. Aminopterin, an antifolic therapy synthesized by Lederle Laboratories and clinically tested by Farber, was the first chemotherapeutic agent to temporarily, but consistently, alter the deadly course of acute leukemia.[12] In her letter to John Gunther, Angela Burns described how Farber's key discovery, ongoing chemotherapeutic research program, and clinical facilities directly affected the course of her daughter's illness. After Mary Sheila's diagnosis, the family physician recommended that they immediately take her to Boston to receive treatment under Farber's supervision. At the hospital, Mary Sheila received blood transfusions and began an aminopterin regimen. The Burnses sought at least a short reprieve for their daughter—a period of health that they hoped would last until the discovery of another new treatment or, ideally, a permanent cure. "In our case," Burns reflected, "there was nothing to lose, and perhaps, miraculously, everything to gain."[13] The Burns family celebrated the dramatic transformation in Mary Sheila's health caused by aminopterin. The drug induced temporary remissions that typically lasted from a few days to several months. Mary Sheila experienced a two- to three-week remission. Although this was only a middling result, her mother wrote joyfully, "She was as well and blooming and normal as she had ever been."[14]

Children cycled from illness, through short periods of remission, and then, inevitably, relapsed. Clinical photographs taken of patients with acute leukemia graphically demonstrated the dreadful manifestations of the disease and its attendant toxic therapies. A long list of complications included: oral ulcers, leukemic infiltration of the eyes, chronic nosebleeds, uncontrollable hemorrhaging, and life-threatening infections that preyed on children's suppressed immune systems.[15] These conditions debilitated sufferers and necessitated frequent trips to and from the clinic as well as extended hospital stays.

Mary Sheila's condition changed constantly, giving her parents and physicians little relief. The chemotherapy regimen lowered Mary Sheila's white blood cell count and reduced the swelling in her spleen and liver but left her vulnerable to dangerous infections. Daily nausea made her stop eating and, consequently, lose weight. In response, her treatment was stopped so that the nausea would abate. Each time the aminopterin was halted, her leukemia—measured quantitatively through regular blood counts—resurged and additional transfusions were administered. Physicians had to maintain a careful balance between aggressively treating the acute leukemia and limiting its toxic effects—infection, nausea, or bleeding—that could also produce fatal consequences.

In August, only three months after her diagnosis, Mary Sheila's health slowly began to deteriorate. Her mother described her daughter's acute leukemia as a "monstrous disease eating her away" during the fall of 1948.[16] Burns's letter painted a horrific picture of the physical toll of the treatment and of the disease itself. Mary Sheila began to experience uncontrollable nosebleeds. Her bleeding episodes became increasingly frequent and progressed to her mouth, gums, and intestines, all sites of the rapid cell division that aminopterin targeted. The capillaries in her legs began to deteriorate, the small vessels in her entire body weakened, and she was soon covered with bright-colored blemishes from the leaking blood. Throughout her illness, Mary Sheila experienced a dozen severe hemorrhages that threatened her vital organs. In January 1949, she died from uremic poisoning, a condition that occurs when elements that are usually excreted in the urine accumulate in the body and produce a toxic state.

Observing her daughter's health fluctuate between periods of health and sickness was agonizing for Burns, yet she conveyed a sense of satisfaction that they had exhausted the available treatment options, writing candidly, "We did everything humanly possible to save our child's life. Nothing was overlooked, no chance was ignored. Many times, our doctor told us it was curtains this time, and she fooled them."[17] The Burnses aggressively pursued treatment alternatives up to the day of their daughter's death. Physicians had planned a direct transfusion of fresh blood between Mary Sheila and her father as the donor. Burns wrote, "Based on her high white blood cell count and low hemoglobin levels, they wanted to give her a complete transfusion—drawing off all of her poisoned blood and pumping in 5000 c.c.'s of fresh blood."[18] Her condition, though, was too precarious to attempt the procedure, so they elected to give her extra blood instead of completely replacing her supply. That night, she slipped into a coma and died less than an hour later. At the time of death, a needle remained inserted into a superficial vein in her ankle, awaiting the partial transfusion.

Children, Hospitals, and Total Care

Mary Sheila spent the months between her diagnosis and death traveling between her home and the Children's Cancer Research Foundation's Jimmy Fund Clinic in Boston. Between 1947 and 1957, the clinic treated 800 children with acute leukemia using Sidney Farber's "total care" concept. According to Farber, total care included, "the application of all the techniques of medicine and surgery for the comfort, well-being and prolongation of life of the child with advanced cancer" as well as "attention to the mental peace of the family as well as to their social and economic problems."[19]

Farber, however, had not pioneered this idea of comprehensive service. Richard Cabot, a noted physician at Massachusetts General Hospital, introduced the concept of adding an in-house social service department to the hospital, and in 1909 a full-time, paid social worker joined the doctor, educator, psychologist, minister, and patient in the "team-work" of the medical dispensary.[20] The social worker visited the home and addressed patients' economic, mental or moral needs. In *Social Work: Essays on the Meeting Ground of Doctor and Social Worker,* published a decade later, Cabot claimed that about 200 other hospitals in the United States had added social work programs. By including the social work professional, Cabot insisted that it helped "retain the individuality of the patient" and "conquer dehumanization."[21] In 1952, Farber adhered to this long tradition, hiring Antoinette L. Peironi as full-time social worker. She developed a social services office at the Jimmy Fund Clinic.

The furnishings and structure of the Jimmy Fund building facilitated this patient-centered system of care. Since the mid-1850s, a number of specialized children's hospitals had evolved.[22] First conceived of as a haven for the urban poor, by the early decades of the twentieth century they had become facilities for acutely ill youth of all ages and classes. The recognition of pediatrics as a medical specialty helped bridge these two phases. The modern spaces and amenities available at the Jimmy Fund Clinic was yet another step in the diagnosis and treatment of sick children. The clinic's waiting room contained a television set and toys to entertain the young patients, painted murals depicting images from familiar children's stories adorned the walls, a merry-go-round provided distraction before appointments, and tricycles provided kid-sized transportation from the waiting from to the laboratory for blood work. To expedite the transfer of new therapies from bench to bedside, the treatment clinic and research facilities were both housed within the Jimmy Fund building. Farber tightly joined the activities of each staff and encouraged the administration of chemical agents in the clinic directly after they had been studied in the laboratory.[23]

Strategically careful to downplay the fluid boundary between experimentation and treatment in the clinic, Farber stated, "[There] is no research, in the popular sense of the term, conducted on the patient with advanced cancer."[24] Notably, this statement resembled C. P. Rhoads's descriptions in *Reader's Digest* about the close link between research and patient care at Memorial. Rhoads declared that the young patients suffering from leukemia supplied "some of the qualities of a wartime emergency" and likened their treatment to the cooperative efforts within the medical division of the Army's Chemical Warfare Service during the war as a way to dramatize the need for rapid advances.[25] He tried to assure the public, however, that patients treated at Memorial were not experimental subjects, but the *voluntary* recipients of promising new treatments developed by investigators at the Sloan-Kettering Institute. He wrote, "Virtually all patients beyond the help of surgery are willing to have new treatments tried on them."[26] Like Rhoads, Farber stated, "We strongly emphasize our research program to parents. Perhaps the benefits of the next medical advance will come in time to help *their* child. We stress our hope for longer periods of remission."[27] Farber and Rhoads recognized popular fears of medical experimentation, but they tried to distinguish their program of rapidly transferring drugs from bench to bedside from any associations with research held in disrepute. Attempting to dispel concerns, they highlighted patients' willingness to participate and their great need for effective agents as justification for their practices. Dire childhood cancer prognoses and the physical proximity of research and treatment buildings at Memorial and the Jimmy Fund Clinic facilitated this simple transfer from animal model to human patient.

As children's cancer care entered the hospital ward and outpatient clinic, nurses, social workers, psychologists, and recreational therapists were needed to provide care for critically ill patients and their families. A handbook for nurses published by the American Society for the Control of Cancer in 1940 contained no information about the education, research, or treatment of the care of children with cancer, but literature on acute leukemia and nurses' integral role in childhood cancer sufferers' medical and personal care began to be published with increasing frequency in the 1950s. In an article from *R.N.: A Journal for Nurses,* Farber attributed "shorter exacerbation periods and longer remissions among children with acute leukemia" to "good nursing care and new anti-leukemic drugs."[28] As children began to live longer and require repeated admissions to the hospital, B. A. Crawford, the nursing supervisor at the clinic, described how nurses' duties expanded. Nurses carried out medical procedures such as sternal marrow punctures and transfusions, watched for possible side effects, regulated diet, and were responsible for the day-to-day hygienic care and discipline of children. They also addressed the emotional and mental needs of the patient and family. Crawford summarized the nurses' responsibilities into one major goal: to eliminate fear of the illness and its treatment.

At the Jimmy Fund Clinic, longer, flexible hospital visiting hours enabled parents to remain present during painful or frightening procedures and help reduce children's anxieties. An on-site penthouse also allowed two or three sets of parents with children in serious condition to stay overnight in the medical facility. Farber considered parents key—even central—members of the "total care" team. Such flexibility was still rare in the care of children and even more so for adults. And, the extra allowances did have pitfalls. The close triad between patients, parents, and nurses could create tension regarding the division of medical and personal caretaking duties. "Mothers are apt to resent having others care for their children," wrote nursing student Mary Brodish. "The nurse may often avoid this resentment by working with the mother rather than in supposed competition against her."[29] Nurses at the Jimmy Fund Clinic taught parents to involve them and to impart skills they would need when their children returned home for weeks or months during periods of remission. Mouth ulcers and sore, bleeding gums posed a particular challenge for maintaining health and hygiene. Parents brushed teeth and gums with a cotton-tipped swab and wiped dried blood from their child's lips and noses.[30] Parents also practiced keeping a careful intake and output record for fluid balance, watching for signs of dangerous transfusion reactions, and applying pressure to intravenous sites to prevent life-threatening hemorrhaging episodes. As Brodish noted, however, "the nurse can decide when situations warrant the parent's participation and when it would be better, either for the parent or the child, for the parent to leave temporarily."[31]

During the 1950s, psychologists and sociologists began to study the immediate and long-term consequences caused by a child's hospital stay.[32] Children's books such as *Johnny Goes to the Hospital* and *Linda Goes to the Hospital* were tools for parents, nurses, or psychologists to introduce children to unfamiliar situations. Simple stories and illustrations explained visits to the doctor, daily hospital routines, and the process of preparing for and undergoing such common surgeries as tonsillectomies.[33] Nevertheless, little could fully prepare children and parents for childhood cancer treatment.

As psychologists and other members of the medical staff began to study the mental and emotional repercussions of childhood cancer, they focused primarily on the children's parents, not the young cancer patients.[34] In the late 1940s, Mary Bozeman and her colleagues in the rehabilitation and psychiatry department at Memorial began a multiphase study of emotional problems faced by the parents of children suffering from leukemia.[35] At request of the pediatric service, they interviewed twenty mothers to evaluate the impact of a fatal childhood illness and to develop new guidelines for care that would help minimize the illness' traumatic effects. Bozeman found that the unpredictability of the disease's course, the pattern of remission and relapse, the relentless side effects of chemotherapy, and the anticipation of their child's inevitable death combined to torment the children's mothers.

In another study that included fathers' perspectives, Beatrix Cobb, a member of the pediatric staff at the University of Texas M. D. Anderson Cancer Center, interviewed twenty parents six months after their child's deaths from acute leukemia.[36] She asked the parents about four different topics: their retrospective reactions to the long-terminal stage of illness; the impact of enforced separation and disruption of routine family life; the impact of illness on the child's well siblings; and the role of religion in their experience of illness.[37] Despite the stresses associated with acute leukemia care, the parents disclosed that they were grateful for the opportunity to spend time with their child as long as his or her illness and suffering were properly controlled. Significantly, Cobb's focus on illness and the whole family circle also extended to other children in the family. Parents had major concerns regarding the well-being of the ill child's siblings when they responded negatively to the sickness and death, for example, losing weight or worrying excessively about their own health.

In one exception, another study focused more directly on young acute leukemia sufferers. The pediatricians wrote, "Hospital practice is increasingly concerned with the management of children with potentially fatal diseases and their families," yet observed, "It is striking to note that so little has been written concerning this aspect of pediatric practice."[38] They found that children felt isolated and depressed during their illness but did not seem openly concerned about death. Still, they cautioned medical professionals that a child's reaction to illness depended on his or her age and cancer type and advised them to evaluate each case individually. To address and assuage both children and their parents, they counseled hospitals to allow parents to serve as partners in the physical care of their children, a system like the one employed at the Jimmy Fund Clinic. The M. D. Anderson model, however, advocated only limited parental involvement. "Because of the considerable anxiety which parents of a child with a malignant disorder faced," they advised it was "undesirable to add to this anxiety by leaving decisions concerning treatment to them."[39] Parental decision making was viewed as guilt inducing, not lending a sense of empowerment or control like physical care. In the 1960s and 1970s, psychological inquiry into patient experience intensified as the development of advanced supportive care, additional chemotherapeutic agents, and effective treatment protocols for acute leukemia ensured years of survival and, in select cases, a cure.

Hormones and 6-Mercaptopurine

Following Farber's discovery of aminopterin in 1947, investigators searched for other agents that would induce partial or complete leukemic remissions, a period after the administration of treatment when the number of white blood cells and other blood cells in the blood and bone marrow were normal (hematological remission) and the patient displayed no signs or symptoms of leukemia (clinical remission). Researchers soon found that two additional groups of drugs induced rapid, temporary remissions in children with acute leukemia—hormones and antimetabolites. In 1949, despite the limited production of cortisone and adrenocorticotropic hormone (ACTH), both hormones were under scrutiny for their possible medical applications.[40] In October 1949, three years after Armour Laboratories first made a preparation of ACTH, researchers gathered at the First Clinical ACTH Conference to give a brief status report. Those assembled represented a variety of research interests and promoted the application of ACTH to patients with gout, nephrosis, rheumatoid arthritis, and other related diseases. Sidney Farber offered the case of a six-year-old boy whose bone marrow was dominated by immature cells called blast forms. The boy complained of pain in his arms and legs, bled from his mouth, and had lost weight. Despite his advanced disease, he had achieved a complete remission with ACTH after nine days of therapy.[41]

In 1950, the availability of ACTH had increased markedly for investigative and therapeutic uses and, in December 1950, Armour sponsored a second meeting. Farber presented the outcomes of seventy-four children treated with either ACTH or cortisone alone or one of the hormones in sequence with folic acid antagonists such as aminopterin.[42] In this investigation, he found that ACTH and cortisone produced the same result. The patients responded in three distinct categories: clinical improvement with hematological improvement, clinical improvement only, and failure. Children who experienced remissions achieved them quickly and improved for between two and thirty-six weeks. The research group observed the optimum results when cortisone or ACTH was administered in sequence with folic acid antagonists. Farber downplayed the harm of "toxic effects" from the therapy, arguing that they could be controlled through measured use of the drugs. In the discussion session following Farber's presentation, physicians from the New England Medical Center in Boston, the Memorial Center for Cancer and Allied Diseases, the Mayo Foundation in Rochester, Minnesota, and others added positive results from their institutions. There was now conclusive evidence that hormones and folic acid antagonists could be used in combination to effectively treat acute leukemia. Farber was confident that these results provided an answer for the frequently heard question, "Should one prolong life in a child with acute leukemia?" The discovery of folic acid antagonists, ACTH, and cortisone gave physicians at research centers the ability to "return the child to a state indistinguishable from normal for a period of weeks or months, and even, in a few instances now, more than 2 years, with the eventual outcome still unchanged."[43] He urged the conference participants to prolong young patient's lives as long as possible, use any available new therapies, prescribe necessary transfusions and antibiotics, and attend to the emotional and social aspects of the disease through a coordinated program of total care.

Less than two years later, experimental results reported by researchers at Memorial Hospital confirmed that the lives of acute leukemia victims were extended by administering hormones and anti–folic acid chemicals consecutively.[44] By using the second agent after cells had become resistant to the first, children lived an average of three to four months longer. The *CA News-*

letter from the same month lauded this form of combination chemotherapy as one of the most intriguing experimental treatment efforts of the time.[45]

Newspaper coverage of these new chemotherapeutic agents differed from the scientific literature, which reported only moderate gains in remission rates and length of survival. Reporters ebulliently described children's journeys from their hometowns to cancer centers for treatment with ACTH. To aid the search for a miracle cure or at least a longer life for these young patients, airlines donated airfare, communities raised funds, and parents accompanied their ill children to specialized treatment and research facilities. "Happy to Be Going Home" declared the caption atop a 1950 photograph of Billy Anderson and Gerald Dunaway, patients at Bellevue Hospital in New York City, waving goodbye to the institution's security guard.[46] Both Billy, an eight-year-old from Summerland, Mississippi, and Gerald, a seven-year-old from Indianapolis, Indiana, had been admitted to the hospital with advanced cases of acute leukemia. In less than two months, physicians had halted the progress of their fatal disease with ACTH and cortisone, calling these drugs the "only medical hope" remaining for these children. Gerald's mother had also turned to her faith, reporting, "We prayed so much I thought the heavens would burst" in hope that spiritual intervention would aid her son.[47] Mrs. Dunaway knew that a miracle would be required to save her son; less than a year before, Gerald's five-year-old sister had been diagnosed and died of leukemia. Mrs. Dunaway urged the reporter to take a close look at Gerald, insisting, "Isn't he the picture of health?"[48] The article ended with the two boys walking hand in hand through the hospital door and "out into the air and light."[49] Although no follow-up story relayed the outcome of their treatment, a short article printed about five-year-old Donna Jean Soderberg only two months later reported the inevitable end to the boys' illness.

In May 1950, the death of Soderberg, the poster girl of the Leukemia Research Foundation, exemplified the brief cycle of treatment and remission associated with these chemotherapeutic agents. Like Gerald and Billy, Donna responded initially to ACTH, was able to return home for Easter, and lived almost a year after her initial diagnosis. After she became resistant to ACTH, physicians attempted to use the hormone a second time, but she did not show signs of improvement and died shortly thereafter.[50] These children's journeys personalized the search for chemotherapeutic agents and lent a sense of hope and urgency to research related to acute leukemia.

The research and development of new drugs and pathways spanned the university medical center, research hospital, government facility, and pharmaceutical industry laboratory. In some cases, collaboration between investigators yielded the most promising results. In 1953, several scientific publications described 6-mercaptopurine (6-MP), an antimetabolite that interfered with cancer cells' ability to manufacture nucleic acids and continue cell reproduction.[51] Developed in a joint research program between biochemists George H. Hitchings and Gertrude B. Elion of the Wellcome Research Laboratories in Tuckahoe, New York, the Southern Research Institute of Birmingham, Alabama, and Sloan-Kettering Institute, preliminary analyses showed that 6-MP was active in some children who were refractory. The term "refractory" designated those patients who failed to go into remission following treatment.[52]

In a 1953 clinical evaluation of 6-MP at Memorial, of eighty-seven children with acute leukemia, forty-one had clinical and hematologic remissions, sixteen experienced partial remissions, and in thirty children the agent was considered a failure.[53] Remissions lasted from one to ten months and researchers found that children were still responsive to other drugs after relapse. Thus, 6-MP could be incorporated into a plan of sequential therapy. The researchers concluded that young acute leukemia patients should receive antimetabolites such as amethoperin and 6-MP first, while cortisone and ACTH should be reserved for emergency situations when a faster acting agent was needed or when the patient became resistant to the initial treatment.[54] The research demonstrated that with no treatment 5 percent of patients lived a year or longer, with cortisone and amethoperin 29 percent survived for the same period, and by staggering amethopterin, cortisone, and 6-MP, the figure rose to 52 percent.[55]

The media covered these developments with keen interest. A science journalist in the *New York Times* boasted, "It looks as if 6-MP is the most useful compound ever discovered for the treatment of leukemia."[56] At a meeting of the American Pharmaceutical Manufacturers' Association, C. P. Rhoads expressed confidence that by refining 6-MP a more specific antileukemic agent could be produced, saying, "Following these principles, it will be surprising if means of cancer control are not found in the foreseeable future."[57] The rapid development of effective chemotherapeutic agents in the laboratory and the cancer clinic led to enthusiastic claims by journalists and investigators alike, but uncertainty remained whether these promises would be fulfilled. And, some asked, at what price to young patients and their beleaguered families?

National Cancer Institute, Cancer Chemotherapy, and Cooperative Groups

In the early 1950s, only a fraction of leukemia patients responded to therapy, and the short, temporary remissions did not significantly extend patients' lives. The announcement of each new agent, however, raised new enthusiasm for this small area of cancer research. The screening program for promising anticancer drugs at the Sloan-Kettering Institute—already responsible for screening about 75 percent of the 2,000 substances tested in the United States—was unable to accommodate the testing of additional compounds supplied by industrial sources.[58] In response, pressure for an expanded program grew among a body of interested investigators and gained force in Congress.[59] A series of special meetings between 1952 and 1954 evaluated the necessity for an expanded chemotherapy program, the best screening methods, an efficient structure for the program, and its financial requirements.[60] The conclusions from these discussions were presented to congressional appropriations committees by National Cancer Institute leaders and other expert witnesses.

Sidney Farber extended his influence beyond the walls of his Jimmy Fund Clinic by entering the national political debate over research priorities and funding. According to a colleague, Farber tirelessly promoted the idea that "the child was the father of the man." In his testimony, he graphically demonstrated that clinical advances in cancer research commonly occurred in pediatrics first and were then translated into similar treatments for adults. Flanked by Mary Lasker and Rhoads and armed with dramatic photographs of a child with leukemia before and after treatment, Farber asked Congress to allocate money for leukemia research. Congress responded this request by appropriating a million dollars to the cause and proposing a new approach to chemotherapy research. Congressional appropriations for new facilities and staff members devoted to cancer, especially chemotherapeutic research and clinical cancer trials, insured a secure place for leukemia and, consequently, all cancers on the national health agenda throughout the 1950s and 1960s. During this period, the annual budget for the Bethesda, Maryland–based National Cancer Institute increased more than $100 million.

In April 1955, the legislative body asked officials at the National Cancer Institute to manage a directed, comprehensive research program targeting acute leukemia.[61] National Cancer Institute was charged with defining the program's research goals and then dividing the tasks into separate contracts for external laboratory and clinical groups housed in universities, at research hospitals, and in the pharmaceutical industry. In *Science News Letter,* the organizational structure of the program was compared to "wartime researches that gave us radar, the atom bomb and other winning developments."[62] Estimating that government, private agencies, and industries had spent approximately $12–15 million identifying and screening chemotherapeutic agents for cancer, they hoped to direct and refine the drug identification process.[63] In response, Kenneth Endicott established the Cancer Chemotherapy National Service Center as a support system for chemotherapy research throughout the country—from the acquisition of chemicals to be tested, through bioassay, into clinical trials. At the National Cancer Institute, the program's center, scientists screened thousands of agents. Each year they tested between 35,000 and 40,000 agents, using mouse models to test the drugs' action against three animal tumor systems: Law's leukemia L1210, sarcoma 180, and mammary adenocarcinoma 755.[64] Congress appropriated $5 million to the project in 1955, $19 million in 1956, and by the following year the national program to promote voluntary cooperative research in cancer chemotherapy represented nearly half of the National Cancer Institute's budget.[65] This level of funding and the vast work it supported indicated the scientific and governmental support for chemotherapeutic research. Only four years later, it was reported that eighty-five chemical compounds had exhibited antitumor activity in animal testing and had been administered to cancer patients to determine humans' biological resistance to the drugs and the agents' toxicity.[66] Work under the auspices of the chemotherapy program and the Cancer Chemotherapy National Service Center investigated chemotherapeutic agents for cancers affecting all age groups but achieved some of its most dramatic results in the fast-growing cancers particular to children.

In 1955, Gordon Zubrod, the clinical director of the National Cancer Institute, and his associates James Holland, Emil Frei III, and Emil J. Freireich directed the medicine branch of the center toward therapeutic research projects on acute leukemia.[67] Holland, Frei, and Freireich founded the Eastern Study Group, and the clinical researchers at Memorial Hospital under Joseph Burchenal were organized as the Leukemia Group. After James Holland relocated to Roswell Park Cancer Center in Buffalo, New York, in 1954, Frei, the administrator of the hospital's leukemia service, and his colleagues at the National Cancer Institute continued their cooperative studies under the title Leukemia

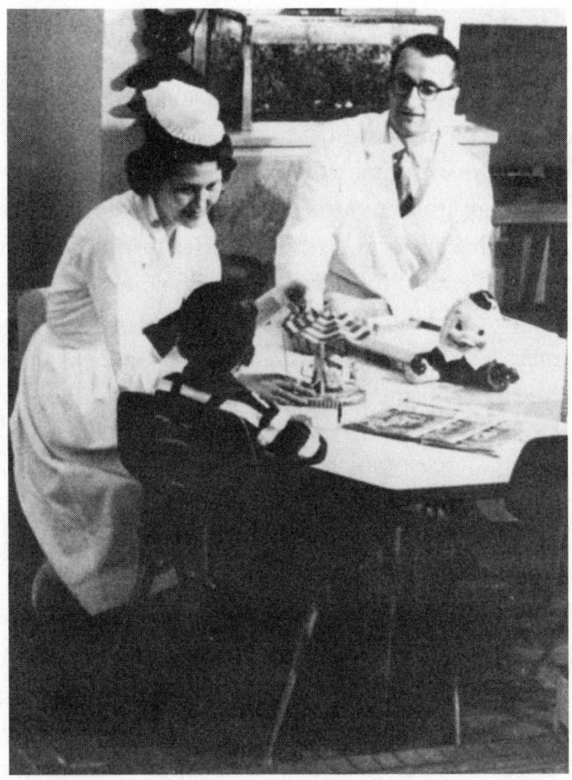

At the National Cancer Institute, clinical research related to pediatric cancers became a very active, yet controversial, part of the program. Despite the resistance that early investigators such as Tom Frei (shown) encountered, child-friendly spaces and special programs to serve this young population and their families grew. Reprinted with permission of the National Library of Medicine.

Group B (later renamed Cancer and Leukemia Group B). The establishment of a formal network enabled the investigators to compare drugs or protocols with very similar remission rates that necessitated a large number of patients in order to detect minute differences. Endicott noted, "A single hospital can rarely make enough observations in . . . highly selected patients to give adequate data in a reasonable time; hence, collaborative research becomes essential."[68] Clinical groups organized patients at the National Cancer Institute Clinical Center, Memorial Sloan-Kettering, M. D. Anderson, the Children's Cancer Research Foundation, and Roswell Park into large-scale, cooperative studies. During their initial study, a comparison of a single and a combination chemotherapy regimen initially studied in a mouse model by National Cancer Institute investigator Lloyd Law, the researcher groups deliberated over proper clinical trial design and statistical analysis.[69]

An addition to Bethesda's facilities furthered clinical research on pediatric acute leukemia patients. Although the National Cancer Institute did not have a pediatrician on staff from 1953 to 1958, children occupied beds in the general clinic. In 1957, the National Institutes of Health added a twenty-bed pediatrics unit to the National Cancer Institute's chemotherapy section for intensive research in cancer in children.[70] Children treated in the large unit ranged from infancy to fifteen years old. Most of the patients only traveled to the unit from the neighboring Washington, D.C., area, but private physicians from across the country also referred their young cancer patients to the facility for further treatment. Children with inoperable Wilm's tumors, osteosarcoma (bone tumors), nasopharyngeal tumors (nose and throat tumors), and neuroblastoma (cancers of the nervous system) were admitted to the unit, but the majority of the patients suffered from acute leukemia and benefited from the unit's emphasis on this area of cancer chemotherapy development.[71] The young patients entered a rigorous research environment at the National Cancer Institute. As Zubrod explained, "Patients are admitted for purposes of total research, not for general care or diagnostic work-up" and in all cases given "the most individualized treatment the unit can offer" during their stay.[72] Under the direction of Frei, a multidisciplinary team of physicians and allied health workers including nurses, dietitians, social workers, and occupational therapists met twice weekly. At the first meeting, the team addressed the case history of each child and the particular problems of the family, including the effects of hospital life and terminal illness on each child and his or her family. At the second meeting, the team and members of the medical staff met to review the clinical findings.

Parents' wish to find new treatments and their desperate hope for a cure sometimes complicated the standardized clinical evaluation of active compounds by specialists. Endicott wrote that maintaining strictly controlled trials was made more difficult by "the urgent need and constant pressure to do something for the hundreds of thousands of patients dying of advanced cancer." "The moment a new drug shows activity in man," he continued, "public and professional pressure is exerted to bring the compound into general use before careful studies can be completed."[73] The authority of the clinical researcher, the integrity of the clinical trial, the risks and benefits to advanced cancer patients, and parents' decisions all had to be constantly negotiated. The clinical studies panel of the Cancer Chemotherapy National Service Center recommended that "the patient's care should be under the complete control of the investigator."[74] Such conflicts over children's care certainly erupted between physicians and parents in hospitals and research centers with recourse to experimental therapies. During a decade when chemotherapy represented a last hope for young acute leukemia victims and their parents, some families encountered unexpected barriers.

Constraints on Care

Angela Burns generally praised her daughter's care at the Jimmy Fund Clinic, but her letter also suggests the toll Mary Sheila's illness and treatment took on the family. She recounted their daily trips from Fall River, Massachusetts, to Boston: "This child made a trip to Boston, 50 miles away, and back, *every day,* and while at clinic, we often had to wait an hour for our turn."[75] In the midst of several trips Mary Sheila's nose had begun to bleed uncontrollably and they had to rush to Boston before she went into shock. The family relied on the Red Cross for a portion of their medical transportation, but as Mary Sheila became frail, she could no longer endure the trip without the comfort of a private car. The Burns struggled to meet the financial demands of her care. Angela Burns wondered how those who traveled from greater distances for treatment could manage, noting, "Everyone must pinch and starve to keep these children with their mothers in a large city like Boston."[76] After Mary Sheila's death, the family established a transportation and hospitalization fund to assist other ill children in the region.[77]

Two other case studies clearly illustrated the role of financial and geographic factors in shaping patients' and families' experiences of acute leukemia in the 1950s. The Jacksons, a working-class family from New York, endured a substantial financial burden during their young son's illness. In his "About New York" regular column in the *New York Times,* Meyer Berger described the Jackson's struggle and New Yorkers' generous response to the medical tragedy.[78] In the second case, the experiences of the Bush family demonstrated how geography impeded some young acute leukemia victims from receiving the latest treatment options, even if that family came from a privileged background. Although the Bush's personal circumstances allowed them to travel from Midland, Texas, to New York City for medical care for their daughter, Robin, few other families living far from noted cancer research centers could have had the resources to pursue similar care for their child.

In 1946, Bill Jackson, like many young G.I.s, returned from duty in the Navy in the South Pacific and married a neighborhood girl. In a few years, the couple had a son, William Sherman Jackson, III, who they nicknamed Skipper. In 1955, his three-year-old son, who was widely known for his lively personality and frenzied pace, abruptly fell ill. The newspaper story explained that a slight slip of his hand while tying his shoe had caused an ugly black eye. Physicians diagnosed acute leukemia. The boy's care—doctor visits, hospitals, tests, and transfusions—drained the family's budget and forced them to borrow heavily to meet the demands of both their household expenses and Skipper's medical care. Although Jackson did not talk openly about his son's illness or the strains on his family at his job as head airfreight agent for American Airlines, the staff at LaGuardia Airport slowly gleaned facts about their trying times. Jackson's coworkers agreed to give a dollar out of each paycheck to help the family and made unsolicited blood donations. Six days later, an article in the *New York Times* announced that Skipper had died in his sleep at the Long Island Jewish Hospital, his maternal grandmother at his side.[79]

Berger's article did not provide detailed information about Skipper's diagnosis or therapy, but it did highlight the demands placed on the family during their only child's illness. Skipper's mother stayed at the hospital full time, and his grandmother provided additional help. The young boy lived only a short time after his diagnosis, but his medical treatment placed an enormous financial burden on the family. From 1943 to 1947, Senator Robert F. Wagner of New York, Senator James Murray of Montana, and Representative John Dingell of Michigan sponsored four bills that supported a national health insurance plan as a component of the Social Security Act.[80] Under this proposal, coverage would have been compulsory, universal, and comprehensive; it would have alleviated the Jackson's financial worries. Under the leadership of Morris Fishbein, editor of the *Journal of the American Medical Association*, however, the AMA vehemently opposed government involvement in payment for medical care. A "socialized" system, he and some of his colleagues maintained, would undermine the entire physician-patient relationship and negatively impact the current practice of medicine. And, there was a second, equally formidable obstacle to overcome. As Jonathan Engel, a scholar of health policy and history, has argued, "Despite the inability of millions to afford necessary medical and hospital services, Americans simply did not want national health insurance." Lacking fundamental support, reformers' campaigns failed.

Although families used a number of alternatives including industrial health plans, prepaid union health plans, mutual hospital plans such as Blue Cross, and private, voluntary medical insurance to help manage their medical costs, those without adequate coverage continued to struggle to pay for their household bills and mounting medical charges. Childhood cancers, like other major catastrophic illnesses, placed an enormous strain on a family budget. Berger's description made it clear that cancer presented a significant hardship for the Jackson family.

Living in a remote area created different burdens on families with children stricken with cancer. Families who lived outside urban areas may not have had access to the experimental therapies available at specialized cancer centers or major research hospitals. An article published in *Southwestern Medicine* advised physicians to learn about the latest chemotherapeutic agents: "The family physician may be called upon to use these products in patients unable to travel to radiological centers because of illness or because of the expense involved."[81] For many patients with acute leukemia, symptomatic treatment and death at home or the local hospital may have been the only suitable options. The Bush family—scions of American politics—confronted some of these challenges when their five-year-old daughter Robin was diagnosed with acute leukemia in 1953.

Former first lady Barbara Bush recalled the obstacles the family faced as they tried to procure medical treatment for Robin.[82] She had scheduled an appointment with the family's pediatrician, Dorothy Wyvell, after her daughter stubbornly insisted that she wanted to remain in bed all day or simply lie in the yard. After analyzing the results from Robin's examination and blood work, Wyvell requested that Barbara and her husband, George, return to her office for a conference. She informed the young parents that their daughter had leukemia. Confronted with the bleak news, they asked the physician for her professional opinion regarding Robin's treatment. In Barbara's recollection, Wyvell encouraged the parents "to tell no one, go home, forget that Robin was sick, make her as comfortable as we could, love her and let her gently slip away. She said this would happen very quickly, in several weeks."[83] Rather than accept this advice, the couple clung to the hope that a second opinion would be different. They asked Wyvell to call John Walker, a relative and physician at Memorial, to inquire whether anything more could be done for their daughter. Walker encouraged them to bring Robin to New York City for immediate treatment by Joseph Burchenal, Lois Murphy, and Charlotte Tan, leading researchers in the field, though he warned them that the therapy would only temporarily extend her life while the search for a cure continued.

Unlike the Jacksons, the Bushes had considerable economic resources. They were able to fly from Texas to New York the next day and admit Robin to the hospital. Both parents divided their time between the Walker home and their daughter's hospital room. The Bushes relied heavily on family and friends for support and the ample blood supply needed for Robin's constant transfusions. Unlike the Jacksons, they also carried medical insurance that would have covered every expense. Memorial Sloan-Kettering (like the Jimmy Fund Clinic) offered free treatment, but the added costs of meals, child care, transportation, and lodging for family members staying near the hospital quickly mounted. They also met parents of other children on the ward who had encountered serious financial problems. When Robin was able to return home for a short stay, Memorial's physicians depended on Wyvell to provide any necessary medical care and Barbara depended on Midland friends for companionship. Bush recalled, "Leukemia was not a well-known disease. Many people thought it was catching and did not let their children get near Robin."[84] Over time, the medicine used to control Robin's leukemia caused side effects that led to a coma. In October 1953, only two months before celebrating her fourth birthday, Robin died. Her parents donated Robin's body to research with the hope that further studies would help lead to a cure for others suffering from this dread disease.[85]

The "One-Boy Whirlwind"

Despite the tragic deaths of Mary Sheila Burns, Skipper Jackson, Robin Bush, and thousands of others, the cancer establishment celebrated gains in the short-term survival of patients with acute leukemia and other cancers. Although few of the American Cancer Society's educational materials directly addressed cancer in children, their fundraising campaigns increasingly used photographs and stories of young cancer sufferers and survivors to encourage donations to the organization's annual drives. In the nineteenth and early twentieth centuries, children had been increasingly used to advertise consumer goods, raise support for child-centered health and welfare reforms, and heighten lay awareness of specific diseases in the popular press and in materials produced and distributed by voluntary health agencies.[86] In the 1940s, March of Dimes campaigns used sentimental images of poster children to direct parental concern toward polio prevention and treatment and to raise funds for their rehabilitation programs.[87] Polio epidemics, not childhood cancer cases, incited the greatest amount of concern in American parents.[88]

Poliomyelitis affected thousands of children across the country as improvements in hygiene and sanitation made children more susceptible to its reach.[89] In an effort to prevent transmission of the disease, parents forbade their children from swimming in public pools, visiting local movie theaters, or playing with neighborhood friends. Leg braces, crutches, iron lungs, and other rehabilitative tools became common symbols of polio's drastic effects. In his study of children's lives during World War II, William Tuttle found that many children had witnessed the effects of the disease firsthand in their classmates and feared polio and the paralysis that often followed more than they feared war. President Franklin Delano Roosevelt, working together with the National Foundation for Infantile Paralysis, wrote that polio eradication was a war goal, "The dread disease that we battle at home, like the enemy we oppose abroad, shows no concern, no pity for the young. It strikes—with its most frequent and devastating force—against children."[90] In keeping with Roosevelt's phrasing, newspaper and magazine stories used war-related terms such as battle, attack, victory, and unconditional surrender to describe polio research programs.[91]

By the late 1940s, John F. Enders, an infectious disease expert at Boston Children's Hospital, had completed preliminary experiments to grow and maintain poliovirus in human tissue cells. From this work, researchers learned that an immunization was needed to protect against three different polio strains. Jonas Salk developed a vaccine and large-scale trials were conducted on children to test its efficacy. In trying to maintain the life and health of child populations, scientists and physicians exposed children "volunteers" to experimental risks.[92] As scientists attempted to translate these results into an effective immunization, American Cancer Society materials began spotlighting a set of lesser-known killers—childhood cancers—to raise funds for education and research. Like polio, a disease widely feared for its crippling effects and periodic, epidemic spread, American Cancer Society images suggested that cancer (often a less visible disease in everyday life) could also harm their children.

In 1953, five cancer patients—two children and three adults—stood on a platform before the American Cancer Society annual fundraising convention as their physicians provided testimony about their cancer treatment. Of the five patients, four-year-old leukemia patient Jennifer McCollum garnered the most attention from the American Cancer Society publicity office. Treated with aminopterin, ACTH, and an unidentified drug provided by Memorial Hospital, Jennifer was promoted as a symbol of the developments in cancer chemotherapy research in the late 1940s and early 1950s. Organizers hoped that her story would inspire those gathered in Chicago to raise the $18 million needed for its 1953–1954 programs.[93]

Two years later, as physician scientists reported on new experimental techniques to the chairpersons and officials at the national meeting, patients who had benefited from the treatments dramatized the potential results. The national meeting in Cleveland, Ohio, was designed to stimulate enthusiasm for the American Cancer Society's April fundraising crusade. At the meeting, two boys afflicted with different cancers demonstrated the current status of cancer research and treatment. Donald Lewis Marteeney from Kansas City, Missouri, had undergone twenty-one surgeries for neuroblastoma before he was two years old, but by the time he was seven, he had been tumor-free for five years and was considered cured. Four-year-old Thomas Nagy of Cleveland was diagnosed with acute leukemia and was treated with cortisone to induce a remission. When he relapsed a few weeks before the meeting, he was treated with a combination of cortisone and aminopterin and experienced a second remission. Farber supplemented the boy's story with slides of patients with leukemia, Hodgkin's disease, and lymphoma who had measurable responses to chemotherapeutic treatment. In their publicity photograph, the boys stood shoulder to shoulder brandishing oversized Swords of Hope while Nagy held a sign printed with the year's slogan, "Strike back at cancer, man's cruelest enemy."[94]

Enthusiasm about the progress made against acute leukemia quickly extended to other childhood cancers that responded to treatment. Throughout the 1950s, children played a key role in motivating fundraisers and articulating the society's educational message. The plight of seven-year-old cancer survivor Leroy Curtis from Denver, Colorado, played a prominent, ongoing role in several cancer campaigns. The story of his diagnosis and treatment was the topic of an article published in Denver's *Rocky Mountain News.* Curtis's journey began when a doctor felt a strange lump in the boy's abdomen at his first checkup. At three months, surgeons removed a malignant tumor near his left kidney. After the operation and ten months of x-ray treatment and hospitalizations, no trace of the cancer remained. To mark his fourth birthday and the probable end of his illness, Leroy had a party with his neighborhood friends, ate cake and chocolate ice cream, and wore a new cowboy outfit.[95]

In 1955 he opened the American Cancer Society's annual crusade by presenting the organization's Sword of Hope to President Dwight Eisenhower. A film of the event was broadcast on television and at movie theaters across the

United States. The chairman of the society's board of directors announced, "We chose Leroy, because we think he is a fine youngster and an excellent example of what can so often be done today to save lives from cancer through early diagnosis and prompt, proper treatment."[96] During a promotional weekend trip to New York City on behalf of the American Cancer Society, Leroy visited the children's ward at Memorial Hospital and joined in a party for the child patients that included two clowns from the Ringling Brothers and Barnum and Bailey Circus. He also appeared on several television shows, toured the Central Park Zoo, and spoke with Meyer Berger about all of his adventures.[97] In 1956, he returned to the national campaign meeting in Cincinnati to pose for pictures and star in the afternoon program led by the year's national campaign chairman, Ed Sullivan.[98] Called the "one-boy whirlwind" in an American Cancer Society article, he led off the show by singing a song while sitting atop the piano. Famous figures, lesser-known cancer survivors, and physicians and scientists participated in the conference and show, but Leroy and his personal experiences with cancer were prominently featured in American Cancer Society publications. Leroy represented the American Cancer Society model of cancer success that included prompt diagnosis, treatment, and cure through conventional medical means. As a survivor of a common childhood tumor that responded to surgery, Leroy and other young children who had been cured of cancer were able to provide a hopeful, sentimental story to motivate fundraising volunteers and inspire benefactors to give generous donations for the research and treatment of all cancers—especially those that required other types of therapies.

Some testimonials, however, were stories that ended in pleas for help. The parents of seven-year-old leukemia sufferer Darya Flagg described their personal experiences in a half-hour radio appeal for the American Cancer Society. From the living room of their farm home, they recorded an unscripted show that was aired nationwide. Darya's grandmother, the principal of Darya's school, their minister, and neighbors also spoke on her behalf. Called "a real-life drama of parents who have refused to give up hope," they mounted a campaign asking listeners to support the cancer crusade and to find a cure for the girl's illness.[99] Unlike Leroy's far-reaching crusade, messages of frank desperation for a cure, not an enthusiastic promotion of the American Cancer Society warning signals, characterized appeals by acute leukemia sufferers and their families who faced inevitable death despite heeding advice about early detection or prompt treatment.

Although American Cancer Society fundraising campaigns broadcast children's stories widely, few educational materials published or approved by the society discussed young cancer sufferers and survivors. This may have been because of the intractable nature of many pediatric cancers. In one exception, *Look* magazine produced an informative booklet on cancer that that was to be widely distributed through companies' employee reading racks. The first article in the booklet featured many children stricken with cancer, the "child killer." In the forward, Charles S. Cameron, medical and scientific director of the American Cancer Society, blamed the rising mortality from cancer on a growing population, an increase in older age brackets, and the growing rate of lung cancer cases but stated that early detection had slowed the rate of increase.[100] He extended his argument to cancer in the young by claiming that raising parents' awareness of the disease would lead to reduced mortality in children—a message repeated from popular articles published in the 1940s.

The examples given in the article, however, challenged the plausibility and benefits of accurate, early detection. Bobby Giampa's parents initially ignored his fatigue, attributing it to his recent entrance into the first grade. After four-month-old Patty Porrine ceaselessly screamed and cried, her mother took her to three doctors before they discovered that she had a tumor in her leg. One physician had dismissed Ms. Porrine's concerns saying, "You modern mothers spoil your children. Just let her cry."[101] Six-month-old Linda Wreith's physician thought that an insect bite may have caused her eye to swell, but when her mother took her to a specialist, x-ray images revealed that several bone tumors had caused the unusual bump. As in the 1940s, experts admonished both parents and private physicians against remaining ignorant of childhood cancers, its symptoms, and the availability of treatment options and cures. Regular physical examinations and constant attention to suspicious pains or lumps were advocated as optimal, though problematic, detection methods.

The *Look* article differed from earlier publications in that it contained a list of therapeutic agents and procedures for treating local and systemic cancers in children. The limited but growing list included: cell "poisons" such as nitrogen mustard, metabolic substitutes such as antifolics and purine and pyrimidine derivatives, experimental therapies using viruses to target and kill cancer cells, hormones such as cortisone and ACTH, and radioactive chemicals to destroy cancerous thyroid cells. Atop the final three pages of the article, phrases proclaimed, "ANGUISH for those who wait for death," "MIRACLES of medicine give parents the will to act," and "HOPE grows that drugs and viruses will stop all cancer." These words and the stories of Bobby, Patty, and Linda demonstrated the dichotomy of helplessness and hopefulness that had characterized childhood cancer in the 1940s and continued to shadow this set of diseases.[102]

The pamphlet also included the rest of Bobby's, Patty's, and Linda's stories. After administering the experimental drug amethopterin, physicians sent Bobby home six days before Christmas in 1951. In July, he returned to Memorial Hospital, where they tried experimental SK 5356 (later renamed 6-mercaptopurine) that permitted him to attend school for four months. He relapsed and died soon after he returned to the hospital for further treatment. Nicknamed the "Hope Child" by her physicians at Memorial Hospital, Patty underwent x-ray treatments to destroy her tumor. Her tumor regressed, but she had to wear a lift on one foot to counteract the side effects of the radiation. When Linda was admitted to St. Luke's Hospital on December 21, 1947, it was the first of thirteen hospital stays for the five-year-old. Surgeons could remove only a portion of her tumors, so physicians tried x-ray therapy and nitrogen mustard to reduce the swelling and control the tumors' growth. The article claimed, "Linda promises to score another victory for chemistry," and her mother added, "I call it a miracle." Despite this hopefulness, a physician at Memorial cautioned, "We never completely discharge a patient."[103] Bobby, Patty, and Linda illustrated how children's experiences with cancer varied widely. All children with acute leukemia died within a year of diagnosis, many children with cancer underwent extensive procedures or experimental chemotherapy treatments, some experienced debilitating side effects from the therapy, and most faced uncertain outcomes.

Experimentation versus Treatment

Angela Burns recognized that her child been in treatment for an incurable disease at the same time she was serving as a research subject for Farber's cancer studies. "Frankly," she wrote, "they were experimenting and they needed children to carry on their work."[104] Research on new chemotherapeutic agents required young acute leukemia patients to endure additional painful procedures and toxic side effects, but parents still sought experimental therapies in hopes of prolonging their child's life and aiding a larger population of young cancer sufferers.[105] When the two-year-old Van Lopik twins from Grand Rapids, Michigan, were diagnosed with acute leukemia, physicians told their parents that there was "no chance" for Eileen Sue's survival and "little chance" to save their daughter Kathleen Jo, who showed signs of leukemia only three months after her sister. Mrs. Van Lopik acknowledged that she and her husband had accepted Eileen Sue's prognosis, saying, "We have given doctors permission to make any type of test they wish or use any attempt at a cure for the benefit of research. Maybe it will help others—even Kathleen."[106] Newspaper articles reported Eileen Sue's death two days later and Kathleen's later in the spring.[107] Statements made by the Burns and Van Lopik families revealed that their decisions required a difficult calculus between the possible risks and benefits for their child.

A series of articles and a letter to the editor in the *New York Times* introduced another example of medical experimentation into a highly visible, public forum. This case involved two distinct populations—children and prisoners—and incited a debate about research ethics among physicians. In June 1949, Louis Boy, a convict in New York's Sing Sing prison, participated in a secret medical experiment. The two physicians involved in the experiment hoped that, by exchanging the prisoner's healthy blood with the leukemic blood of eight-year-old Marcia Slater, they would transfer an unidentified factor that would combat leukemia and, eventually, cure the young girl. They also believed that the prisoner might develop an antileukemic substance during the exchange that could then be used to fight leukemia in other patients. Physicians directly transferred half a pint of affected blood from the ill child to the veins of the convict through a piece of rubber tubing. They then gave the child a pint of his blood. The exchange continued for five hours a day on four consecutive days until approximately eighteen pints had been exchanged between Boy and Slater.[108] Physicians regularly used transfusions to strengthen leukemia patients and to make them more comfortable, but newspaper articles describing the experiment claimed that the recipient's blood had never been added back to the donor's veins. After the extensive transfusions, the prisoner remained under guard in a ward at the Ossining, New York, hospital while the girl rested in a private room. By the end of the week, initial tests on the prisoner recorded no abnormalities in his blood and bone marrow and a small improvement in the young cancer patient. The experiment, a last resort for Slater, did not succeed in saving the girl's life.[109]

In interviews with the press, physicians involved with the procedure publicly justified the experiment's possible dangers and benefits. Slater's doctor, Harry Wallerstein, a physician at Jewish Memorial Hospital in New York, stated that he conducted this experiment with a prison volunteer because he needed to enroll a subject who was willing to take risks.[110] Although previous experiments of this nature carried out on laboratory animals had shown that it was not possible to transmit the disease through an exchange of blood, an authority declared, "I would not care to have leukemic blood pumped into my own veins."[111] Wallerstein and James B. Murphy, head of the cancer division of the Rockefeller Institute for Medical Research, confirmed that Boy had volunteered without the promise of reward and with full knowledge of the possible fatal risks including a "very, very remote" chance that he could contract leukemia. For the next year, physicians planned to monitor his blood for the presence of leukemic cells. The researchers' statements and subsequent actions suggested a level of uncertainty regarding the experiment's effects on both Slater and Boy.[112]

The Boy-Slater experiment raised moral questions regarding the acceptable levels of risk to research subjects in medical experiments. On June 8, a letter to the *New York Times* from Ludwig Gross, a physician and the chief of cancer research at Veterans' Administration Hospital in New York, fiercely censured the experiment, calling it "open to considerable criticism from both a scientific and also from a purely humanitarian point of view."[113] Likening leukemia to a death sentence that could only be temporarily delayed by new chemotherapeutic agents and radioisotopes, he asked whether the prisoner was properly informed of the dangers that would result if he contracted this disease during the transfer. Unlike Wallerstein and Murphy, Gross also claimed that similar transfusion experiments had been conducted previously between leukemia sufferers and those with terminal illnesses. Gross was outraged by the vast amounts of infected blood that were transferred to a healthy human volunteer. He condemned the experiment, saying that Slater would gain only the temporary improvement she would have received from a standard transfusion, a small chance remained that an experimental accident could also kill Boy, and the transfusion had few significant benefits to science or medicine. Test results from Boy's final medical examination at Jewish Memorial Hospital showed that Boy had not been harmed in the experiment.[114]

Why did the Slater-Boy experiment stimulate such a volatile response? In his editorial, Gross focused on issues surrounding Boy, the healthy prisoner "volunteer," but also briefly addressed the considerations required when experimenting on a child—specifically, a child suffering from a terminal illness. By the time of this debate, there had already been a long history of children serving as research subjects.[115] In the early twentieth century, accusations of "human vivisection" and the use of children as "guinea pigs" prompted debate over the limits of acceptable human experimentation. The success of medical research in the 1930s and the benign picture of investigation (that physicians used only their own bodies and those of volunteers) obscured the more complex realities of the research activities in the 1940s.[116] Physicians involved in this experiment and in all clinical research on acute leukemia had to meet two interconnected, yet often opposing, goals—the pursuit of new medical knowledge through experimentation and the responsibility to protect children from harm.

Despite the creation of the Nuremberg Code, a document composed of ten key ethical principles that was drafted in response to Nazi medical experiments, there continued to be little consensus among American researchers of the limits of medical research.[117] The use of dying children reflected an uneasy, constantly negotiated union of experimentation and therapy. At a time when few formal guidelines or professional criticism restricted institutional human experiments, several factors may have spurred Gross's reaction: the child participant, her particular disease, few therapeutic benefits, or the direct blood transfusion from a male prisoner to a young girl. The experiences of Mary Sheila and Marcia Slater poignantly demonstrate that the overlap between childhood cancer research and treatment merits a careful scrutiny, both to expose past abuses and to illuminate present and future concerns of patients, parents, and practitioners.

In 1952, Farber boasted, "If the problem of resistance, either initial or acquired, of the leukemic cell to the folic acid antagonist could be solved, the usefulness of the antagonists in acute leukemia could be compared with justice to that of insulin in diabetes."[118] He repeated this belief again four years later, expanding the problem of resistance to all antileukemic agents developed during the subsequent four-year period.[119]

This obstacle prevented the cure for acute leukemia promised by select physicians, promoted by the popular media, and expected by the public; however, children with cancer were living longer. In the 1940s, fewer than 5 percent of children with acute leukemia survived one year following their diagnosis. By 1956, more than half of the children receiving treatment for acute leukemia survived a year from clinical diagnosis of the disease because of the aggressive use of chemical agents and the addition of adjunct therapy such as transfusions of whole blood, antibiotics, and maintenance of fluid and electrolytes.[120] Yet, these changing forms of medical management created new challenges for physicians and allied health practitioners, children, and parents. By the mid-1950s, some children were able to take their pills at home and come to the outpatient clinic only once or twice a week for general examinations and bone marrow aspirations to monitor their leukemic cell counts.[121] This relieved physicians and nurses of the child's daily care, but parents were now charged with administering medications, checking for signs of toxicity, stopping a dangerous hemorrhage, guarding against infections, and shuttling the child to frequent medical appointments. These responsibilities multiplied as children neared the end of their lives. In the next decade, parents' duties continued to expand as two collaborative groups—Leukemia Group B and the Acute Leukemia Task Force—produced consistent, lengthy remissions for most children and provided proof that chemotherapy could cure acute leukemia.

CHAPTER FIVE

“Who’s Afraid of Death on the Leukemia Ward?”

REMISSION, RELAPSE, AND CHILD DEATH IN THE 1960S AND 1970S

In 1961, Peter De Vries, a prolific writer and contributor to the *New Yorker,* published *The Blood of the Lamb,* a novel that documented the tragic illness and death of a young girl from acute leukemia.[1] In the first half of the book, Don Wanderhope, the book’s protagonist, recounted the sickness and suffering of close family members and friends. After this long series of events, his eleven-year-old daughter, Carol, was diagnosed with acute leukemia. The second half of the novel offered an exacting description of his daughter’s illness and Wanderhope’s own inner struggle to comprehend a disease he termed “the Slaughter of Innocents” against his beliefs about religion and medical science. *The Blood of the Lamb* drew criticism for its radical departure from the author’s comedic style, a style that was defined by its wise use of its puns and malapropisms. De Vries’s quick wit and dark humor allowed him to make biting evaluations of the contemporary American scene and the absurdities of modern life. This unusual voice caught the attention of critics and readers alike. The dark tone and heartrending plot of *The Blood of the Lamb* led the reviewer from the magazine *Commonweal* to write, “The grief of Mr. De Vries’s hero is agonizingly real, as real as life.”[2] Another predicted, “Those who have laughed with him in the past and during this book will not begrudge him their tears.”[3] Far from a purely fictional account, De Vries had based the book closely on autobiographical events and used the writing process to cope with the tragic death of his ten-year-old daughter, Emily, from acute leukemia.

Close parallels existed between the events described in *The Blood of the Lamb* and patients’ experience of acute leukemia in the 1960s. Despite improvements in the survival rates of children with retinoblastoma, Wilm’s tumor, and other common childhood tumors, leukemia remained a frustrating enigma. Chemotherapeutic agents delayed, but did not deny, death early in this decade, and longer survival through chemical means brought a list of benefits and challenges to physicians, parents, and children: complicated multidrug treatment regimens, severe physical side effects, and the inevitability of death once experimental chemotherapeutic options had been exhausted. Children relapsed when cells became resistant to a chemotherapeutic agent and the bone marrow again became packed with diseased cells. Physicians administered another drug, if available, but if the limited treatment alternatives had been exhausted, the child died. The new pattern of illness fostered debates among medical professionals at conferences and in medical journals as the coordination between hospital, outpatient, and home care was negotiated. This chapter explores the contests that erupted over each phase of treatment but also attempts to capture the personal, variable component of disease experience, specifically acute leukemia in the 1960s.

The Blood of the Lamb

De Vries's family life and local community became the basis for many of his writings.[4] In 1956, his daughter Emily was pictured with her father on the dust jacket of *Comfort Me with Apples* and the novel was dedicated "To Emily With Love."[5] She died shortly after, on September 19, 1960. De Vries's own stay at a church-sponsored tuberculosis sanatorium in Denver, his father's debilitating mental illness, Emily's premature death, and his own experiences during her illness served as the basis for the events in *The Blood of the Lamb*.[6] Although Carol's birth did not take place until the book's halfway point, De Vries's biographer J. H. Bowden argued, "It is her death—impending and then actual—which informs the book, shapes it."[7] Wanderhope's character provided readers with an intimate exploration of two themes: Carol's diagnosis, illness, and death and his intense theological inquiry into the limits of humanity, God and religion, and scientific medicine. Though not an exact description of De Vries's experiences with Emily's death, the biographical similarities were striking. In a 1967 manuscript documenting the history of the Department of Pediatrics at Memorial Hospital for Cancer and Allied Diseases, pediatrician Harold Dargeon wrote, "Although, we did not consciously contribute to the book published by the well-known author Peter De Vries, his own experiences at the hospital are a matter of record in the volume entitled *Blood of the Lamb*."[8] "As would be expected," his bibliographer noted, "this disaster needed to be organized and exorcised fictionally by De Vries."[9]

De Vries's work, in the form of published poems, novels, essays, plays, and reviews, reached wide audiences through a variety of noted literary publications and anthologies. A lengthy summary of *The Blood of the Lamb* was reprinted by *Reader's Digest* Condensed Books and titled "Carol."[10] This abbreviated edition was translated into eight languages, and the original book was reprinted as a "Popular Library" volume. Despite the fictionalized nature of the personalities and events in De Vries's personal experiences, *Carol* provided readers with a provocative character and a substantial window through which one could gain insight into the deep impact of acute leukemia on the child and family in the 1960s.

Recommendations and Reality

Like many children with acute leukemia, Carol was not immediately diagnosed with the disease. She first fell ill with a mild fever and recurrent back pain. A prescription of antibiotics failed to relieve Carol's common flulike symptoms, and she was admitted to the hospital for x-rays, blood tests, and a throat culture. Doctors detected and treated a streptococcus infection, but a second episode of fatigue, fever, and aches prompted her physician to order another blood test and a bone marrow aspiration. The pathology report established a definitive diagnosis of acute leukemia.

A flurry of popular articles written about childhood cancers in the 1960s in *Family Circle, Reader's Digest, Good Housekeeping,* and other publications repeated the messages promoted by earlier articles alerting parents to common cancer sites in children, warning them about suspicious symptoms, promoting early detection, urging them to promptly seek physicians' counsel, and updating readers about advances in treatment.[11] They also began promoting yearly physical exams for preschool-age children as vital tools for prompt tumor detection and treatment.[12] In *Health,* a publication of the American Osteopathic Association, the author noted that children experienced a high cancer rate during this period that coincided with a gap in medical care that occurred between the first-year checkups and the required school entrance medical exams. "During the first year of life," Stewart wrote, "infants keep regular doctor appointments, by and large, transported back and forth by conscientious modern mothers who are properly respectful of the widely publicized diphtheria-whooping cough-tetanus and polio immunizations."[13] In *Reader's Digest,* a distressing story urged parents with a family history of the eye tumor retinoblastoma to watch closely for physical signs of the disease such as a white speck in the eye or an unusual widening of the pupil.[14] Studies of twins and the tumor's incidence had proven that retinoblastoma was caused, in large part, by heredity. The article described how one watchful mother detected the growing tumor in each of her three children and saved the lives of two of them through her vigilance. Unlike the Vasko and Colan cases, new technology allowed physicians to pinpoint the small tumor with radiation and preserve the affected eye. With early detection and new treatment methods, the cure rate for retinoblastoma had improved from a fatal outlook to a 90 percent cure rate by 1964.[15] The message of joint responsibility by the child's mother and pediatrician for cancer detection continued into the 1960s, but it focused on treatable childhood tumors. Reports virtually ignored acute leukemia because without curative agents, early detection did not equal longer survival. Instead, articles that included information on acute leukemia outlined incremental steps made in chemotherapy research.

Chemotherapy and Miraculous Cures

In 1963, *Cosmopolitan* magazine featured a special report on cancer in children that vividly extolled the progress in the treatment of acute leukemia and other common childhood cancers. Notably, it also recounted parents' concerns about cancer causation and informed readers that the number of such inquiries had risen sharply. Many parents who brought their children to the Children's Cancer Research Foundation wondered whether the fallout from nuclear testing had contributed to their child's cancer, fearing that fission products of uranium and plutonium like strontium-90 and uranium-131 had tainted their family's milk and food. They had heard the message that isotopes could lodge in children's developing tissues and slowly accumulate over time, wreaking genetic damage and causing cancers, as these dire warnings had been disseminated through a number of highly visible means: congressional hearings, popular magazine articles, and boycotts, protests, and advertisements sponsored by Women Strike for Peace and the National Committee for a Sane Nuclear Policy, a peace organization cochaired by pediatrician Benjamin Spock.[16] Now, with few concrete answers as to why their child was sick with a rare, fast-growing disease, they searched for a plausible explanation; however, two major factors—a variable delay between radiation exposure and diagnosis of the disease and uncertainty about "safe" levels of radiation in children—made it difficult to definitively make a cause-and-effect relationship. Thus, the article focused primarily on a more optimistic angle.

Highlighting Farber, the Children's Cancer Research Foundation, and the "developing science of chemotherapy," the article acknowledged the dreaded nature of common childhood cancers but boasted of Farber's ability to cure cancer through new chemotherapeutic agents.[17] Wilms's tumor, in particular, responded dramatically to chemotherapy. In medical lectures, Farber showed a slide of a boy at age two and a half with Wilms's tumor and pulmonary metastases. He began, "Considering the size of the tumor and the presence of cancer in the lungs, the doctor treating this boy regarded an operation as contraindicated, and told the parents the outlook was hopeless."[18] He went on to contrast this fatalistic attitude with the approach toward the boy's condition at the Children's Cancer Research Foundation: the boy was immediately scheduled for radiotherapy appointment and given the antibiotic actinomycin D. "In a few weeks," Farber claimed, "his lungs were clear of cancer, and the tumor had reduced in size to the point where surgery was possible."[19] The next slide showed the boy at age seven with no evidence of cancer. New treatments for Wilm's tumor had transformed its cure rate from 20 to 40 percent with surgery and postoperative radiotherapy to 80 percent when the antibiotic actinomycin D was added. Farber assumed that a similar radical improvement would take place for acute leukemia sufferers in the near future. As in the late 1940s and 1950s, Farber sought "not only to prolong life, but to prolong good life" as long as possible in the event that a new, effective drug was discovered.[20]

Official American Cancer Society publications reinforced this optimism. In *The Truth about Cancer,* Charles S. Cameron, former medical and scientific director of the American Cancer Society, mimicked the dramatic, perhaps exaggerated style in newspaper and popular magazine articles to bring attention to potential power of chemotherapeutic agents against acute leukemia. Cameron told of the suffering of a leukemic child who was brought to the hospital on Labor Day. He characterized the young girl as "pale, drawn, distressed by constant bleeding from gums and nose, and with a prodigiously swollen tummy, unable to eat or even to walk, facing, by standards of five years ago, a few more weeks of increasing misery and ebbing life."[21] Drawing upon the biblical story of Christ's miraculous resurrection from the tomb, Cameron framed his story as a medical miracle produced by chemical compounds. He described the young girl's remission: "I have seen the same child Easter Sunday afternoon, gathering the newly blossomed daffodils in her mother's garden, jumping rope with her friends, to all outward appearances a healthy, happy child."[22] This pair of observations convinced him that something "exciting and important has happened" in cancer chemotherapy research and that the experiences of this little girl indicated "only the beginning."[23] As a promoter of the American Cancer Society message of hope, Cameron used the story to assure readers that the advent of additional drugs held the potential to deliver more consistent, prolonged results that would ensure a rebirth for terminally ill cancer sufferers. De Vries's novel also used religious imagery but used it to deliver a different lesson. The symbolism and language signified the death of an innocent child from acute leukemia and, perhaps, the sacrifice of one child to improve the health of other sufferers. Through the thoughts and voice of Don Wanderhope, sections of the novel exposed a father's considerable wariness about experimental cancer treatment, his daughter's suffering, and the utility of continuing the aggressive treatment.

In *Blood of the Lamb,* Dr. Cameron, the Wanderhope's family physician, visited their home in Westchester County to deliver the grave diagnosis personally. He advised Wanderhope to take Carol to Westminster Hospital in New York City for access to the "world's leading authorities," extensive scientific research laboratories, and medical treatment facilities dedicated solely to cancer on childhood cancer.[24] Westminster was modeled after Memorial Hospital, the location of Emily De Vries's acute leukemia treatment. In response to Wanderhope's uninformed queries about possible treatments or cures, Cameron admonished,

> My dear boy, where have you been the last ten years? There are first of all the steroids—cortisone and ACTH—which give a quick remission. The minute she's pulled back to normal with those, Dr. Scoville will switch her to the first of the long-range drugs, some of which he's helped develop himself. If they should wear off, there's—but let's cross those bridges when we come to them.[25]

Such a series of chemotherapeutic agents typified the therapy for acute leukemia therapy in the 1960s. By administering effective drugs sequentially, gradually shorter remissions were linked with the goal of extending the child's life as long as possible. The physician not only made reassuring predictions about the length of Carol's remission but also imparted genuine hope for a cure in the near future, saying, "They're working on it day and night, and they're bound to get it soon."[26] After the diagnosis, Carol's care immediately shifted from Dr. Cameron to the specialized cancer center, where she could receive advanced treatment based on late-breaking research results.

At Westminster, Dr. Scoville, a pediatrician and specialist in childhood cancers, reiterated Cameron's enthusiasm for the potential of chemotherapeutic agents to prolong and, perhaps save, Carol's life. He began by examining her and ordering another battery of laboratory tests to confirm her previous diagnosis. At a private meeting between Scoville and Wanderhope, he admitted that her condition was worsening but listed the possible chemotherapeutic agents that induced or maintained temporary, complete remissions in acute leukemia patients: steroids, methotrexate (a less toxic form of aminopterin), 6-mercaptopurine, and other experimental agents. Although he was intentionally evasive about Carol's chances for a permanent cure, he stated,

> Chemotherapy—drugs—is the scent we're on now, and it's only a few years ago we didn't have anything at all. It's quite a game of wits we're playing with this beast. The 6-MP, for example, breaks the cells up nutritionally by giving them counterfeit doses of the purine they like to gorge themselves on. I hope we'll have some other pranks to play on him soon, and if there are, you may be sure the clinic downstairs will be the first to try them out. There's nothing hot at the moment, but who knows? It's an exciting chase, though I can't expect you to look at it that way at the moment.[27]

For Scoville, the discovery and testing of chemotherapeutic agents was a professional challenge, a kind of intellectual game. Wanderhope appreciated the function of research hospitals, but he was also keenly interested in practical, available chemical agents that could be applied in his daughter's case. Scoville provided hope for acute leukemia patients in the form of scientific and medical discoveries but recognized that he could actually only promise a long series of medical tests, temporary remissions, and an uncertain future for Wanderhope and his daughter.

Scoville represented a member of the small cohort of physicians involved in Leukemia Group B, one of the most successful of the ten collaborative clinical research groups sponsored by the Cancer Chemotherapy National Service Center. Leukemia Group B organized clinical trials and evaluated the activity of cytotoxic drugs, modified chemotherapeutic protocols for children, and managed the dangerous side effects of drugs. As a consequence, they significantly increased the one-year survival rate of acute lymphocytic leukemia.[28] Children survived longer but invariably relapsed and died. In his conversation with Carol's father, Scoville optimistically portrayed the ongoing search for new agents by researchers and steered Wanderhope toward the promise of further chemical breakthroughs for his daughter.

By the early 1960s, Leukemia Group B became a part of a broad cooperative program organized to test chemotherapeutic agents against cancer in clinical settings. New funds, appointments, and a special task force suggested the political importance attached to the development of new chemotherapies. In his analysis of national medical and cancer research policies, Stephen Strickland described the testimony from experts such as Farber that persuaded Congress of the great promise of cancer chemotherapy. He wrote, "In proposing $24 million above the House recommendation for the National Cancer Institute in 1960, the Senate appropriations subcommittee indicated that the chemotherapy program was one area where it wished to see increased activity."[29] In 1960, two leaders in chemotherapy research moved from the Cancer Chemotherapy National Service Center to leading National Cancer Institute posts; Kenneth Endicott became National Cancer Institute's director and C. Gordon Zubrod became the institution's scientific director. In spring of 1962, Zubrod created the Acute Leukemia Task Force, a special group directed "to engineer a cure" of acute leukemia. Drawing upon management principles used by IBM to direct particular projects, the task force used the industrial model "because the problem of the cure of ALL [acute lymphocytic leukemia] was seen as a technical one: five efficient drugs existed already, and researchers were convinced that it would be possible to design drug protocols that would kill all of the residual malignant cells and prevent relapses."[30]

In 1964, a review article by Joseph H. Burchenal of the Sloan-Kettering Institute for Cancer Research and Memorial Hospital summarized the "present armamentarium" of five different classes of chemotherapeutic agents available to treat acute leukemia in children and adults.[31] Between 1948 and the early 1960s, aminopterin (methotrexate), prednisone, 6-MP, cytoxan, and vincristine were shown to be effective against acute leukemia in children. Patient records at Memorial Hospital through 1963 show that children treated with the nitrogen mustards alone in 1946–1947 survived an average of two months, whereas those treated with the full set of agents in the early 1960s survived an average of thirteen months from the start of treatment. In the mid-1960s, preliminary studies demonstrated that the new agents cytosine arabinoside and daunomycin were also active against experimental tumors, childhood leukemias, and forms of adult leukemia.[32] Treated in 1960, Carol Wanderhope did not have recourse to the complete "armamentarium" of agents or combination protocols, but she did receive steroids, 6-MP, methotrexate, and an unnamed, experimental drug. New chemotherapeutic treatments for acute leukemia and other common childhood tumors again raised hopes for a cure, but deaths continued to be more common than cures in this set of cancers.

One branch of the Acute Leukemia Task Force focused on modifying dosage schedules to maximize efficacy while minimizing toxicity. M. C. Li, a National Cancer Institute researcher, had demonstrated that an equally effective, but less toxic form of aminopterin named methotrexate cured choriocarcinoma, a rare reproductive cancer. Choriocarcinoma studies in the L1210 mouse leukemia model showed that methotrexate showed more activity when administered intermittently than with the daily doses given in standard treatment.[33] As a result of this work, Leukemia Group B developed a similar dosing pattern in humans in order to lengthen remissions. Their clinical experiments compared the standard oral daily dose with intravenous administration, an intermittent schedule, and massive doses. Researchers also launched the first quantitative "adjuvant" study to probe whether administering an active agent during complete remission lengthened the period before relapse significantly.[34] This practice, termed adjuvant therapy, became the focus of many cancer research programs.

Investigators rigorously compared the administration of single agents or drug sequences to combination therapy protocols. Combination or cyclic therapy seemed to prevent the problems of drug resistance.[35] Based on encouraging results in the L1210 mouse leukemia model, Emil Frei III and Emil J. Freireich predicted that if two chemotherapeutic agents were administered simultaneously patients could respond to either. The researchers used a three-arm design to test whether they should give methotrexate, 6-MP, or both and to quantify the difference between using a single agent or two agents simultaneously. The combination produced a synergistic effect—the number of children in remission equaled the expected rate for both individual agents. If the researcher maintained the full doses, the combination improved the complete remission rate from 20 to 90 percent.[36] Clinical investigations also demonstrated that the best agents for inducing remissions differed from those that were most effective for treatment during remission: vincristine and prednisone were the optimal combination for inducing rapid, complete remissions, but 6-MP and methotrexate substantially lengthened remissions.[37] On the basis of these results, the National Cancer Institute researchers designed the VAMP protocol (vincristine, aminoptherin, 6-mercaptopurine, and prednisone) to measure the activity of four antileukemic agents given simultaneously.[38] After three cycles of VAMP, they stopped treatment to observe its effect on human patients. Calculations based on cell kinetics—a method to predict total cell death mathematically—had predicted a cure. Although "cures" could not be declared immediately, within six months researchers were confident that VAMP was inducing longer remissions than had been observed previously.[39]

National Cancer Institute researchers modified VAMP by creating a new two-cycle drug regimen known by the acronym BIKE (from "bi-cycle") that was initiated after remission was achieved. The physicians used vincristine and prednisone simultaneously to induce remission and then administered a full dose of methotrexate and 6-MP. Then the first two agents were resumed. The next study combined VAMP and BIKE into POMP (prednisone, vincristine [oncovin], methotrexate, 6-MP [purinethol]) which used the same agents as BIKE, but treated patients longer and used higher drug doses.[40] Based on this series of key studies, researchers discovered that the most effective treatment regimen was not continuous low-dose therapy, but an intermittent schedule of aggressive treatment even after all evidence of leukemia had disappeared. This plan induced remissions very quickly, caused less cell toxicity, and allowed for normal cell recovery. New experimental agents and this series of protocols introduced a rapid rise of the multidrug, multicycle regimens that were continually modified and accepted as the standard treatment for acute leukemia and many other cancers found in children and adults from this point forward.

"Everything Was Fine"

Children like Carol Wanderhope who responded positively to chemotherapy entered a cyclical pattern of treatment, remission, and relapse that ended when the supply of effective agents was exhausted and the child died. After discussing Carol's diagnosis, Scoville led Wanderhope and his daughter to the outpatient clinic, the site of Carol's physical exams and procedures once her condition stabilized. Regular bone marrow aspirations would help physicians chart her condition and detect remissions, relapses, and infections while she lived at home. The family also relied on the clinic during unexpected crises. Scoville warned Don Wanderhope that the disease would destroy his daughter's platelets and hinder her blood's ability to clot properly. He supplied him with cotton packing to plug her nostrils during a nosebleed, but before her next scheduled visit, Carol, her father, and the family housekeeper traveled to the clinic for urgent aid when the bleeding became uncontrollable.[41]

The novel chronicled the repeat hospitalizations one young girl required for the management of her disease. After the nosebleed had been stopped, Carol was readmitted to the hospital for a transfusion and a dose of cortisone to induce her first remission more quickly—the first of many emergency visits. She was housed in the Children's Pavilion—a separate place for children described by her father as, "a bedlam of colliding tricycles, bouncing balls, and shouts for nurses and the volunteer workers known as Bluejays, so named for the color of the uniforms in which they bustled about on non-medical errands," at the time of his first visit.[42] Although he dismissed the lively Bluejays as frivolous society women, Wanderhope credited them with contributing to the prevailing attitude that "Everything Was Fine" despite the presence of children with amputated limbs and massive bleeding.[43] A few days later, Scoville was unable to palpate Carol's spleen, a sign that her platelet count had improved through the cortisone treatment, and Carol returned home.

Wanderhope depended on Scoville's advice and Carol's regular outpatient visits to properly monitor her condition. As Carol cycled through a sequence of chemotherapeutic agents, she experienced side effects that signaled the activity of her treatment and, consequently, fluctuations in the severity of the leukemia. The steroid therapy increased Carol's appetite and required her to strictly adhere to a low-salt diet to prevent high blood pressure. She gained a significant amount of weight and became self-conscious both about her new body shape and the way her friends viewed the dramatic change.[44] At her next visit to the clinic, Carol was given 6-MP to help maintain her remission. The first doses of 6-MP raised concerns about a new set of side effects, including mouth sores, vomiting, and diarrhea that indicated the drug's toxicity. Wanderhope reported to Scoville that Carol's gums were painful and inflamed, but he dismissed the father's concerns, saying that it was best if she could remain on the "edge of toxicity" while maintaining a solid remission. Once a sternal puncture, a procedure to sample bone marrow, demonstrated that Carol had stabilized on the 6-MP, she was given a three-week interval before her next appointment. Six months after starting 6-MP therapy, Carol developed resistance to the drug, but she quickly responded to the next drug, methotrexate. After only three months on the new drug, though, Carol complained of chronic headaches and problems with her eyes.

The symptoms suggested that treatments had controlled the leukemia in her blood system, but cells had continued to proliferate in the meninges, an area of the central nervous system where the drugs could not penetrate and a sanctuary for leukemic cells formed. Children like Carol, who experienced long remissions, risked the development of meningeal leukemia because layers of leukemic cells lined the membranes to constrict the movement of spinal fluid. Patients experienced headaches and changes in their vital signs as a result of the added spinal pressure, but they remained systemically in remission. Physicians used a lumbar puncture to detect leukemic cells in the spinal fluid and intrathecal injections to transfer massive doses of methotrexate directly into the space surrounding the spinal cord. This improved children's symptoms for a relatively brief period before relapse occurred.[45]

As Carol's body became resistant to methotrexate, she was hospitalized again for severe bleeding. Wanderhope described the uncomfortable cotton packing that extended from her nose into her throat to stop the blood flow. Carol bore obvious physical signs of her constant, invasive medical treatment, including scars on her hands from intravenous and transfusion needles and on her breast from a marrow aspiration. Using religious symbolism to compare his innocent daughter's suffering to that of the crucified Christ, Wanderhope referred to her injuries as "stigmata." The bleeding, open "stigmata" designated Carol as a sacrifice to the disease and to the goals of medical research. Her father felt helpless as he became a bystander to the unfolding events.

Debating Cancer Centers

Like Emily De Vries and Carol in *The Blood of the Lamb,* many children were immediately referred to cancer centers by their pediatricians.[46] Despite the promotion of these centers by cancer specialists, proceedings from a one-day conference "Care of the Child with Cancer" illustrated the benefits and possible drawbacks of the facilities.[47] The November 1966 meeting was sponsored by the Association for Ambulatory Pediatric Services, an organization for directors of pediatric outpatient departments interested in promoting improved care, teaching, and research in pediatric ambulatory facilities, and the Children's Cancer Study Group A, physicians engaged in the cooperative clinical trials of chemotherapeutic agents under the direction of the National Cancer Institute. Participants critically evaluated the proper role of their home institutions. In addition to internal review and assessment, one convener noted, "We have much conviction but little knowledge about the feelings of the medical consumers and the health professionals outside medical centers."[48]

Most of the participants agreed that specialized centers were the optimal site for treating childhood cancer patients, though there was dissention. Specialists like John R. Hartmann, director of the hematology and oncology division at the Children's Orthopedic Hospital and Medical Center in Seattle and chairman of the Children's Cancer Study Group A argued vehemently that it was worthwhile for pediatricians to refer their cancer patients to specialized cancer centers for advanced care, despite the debilitating effects of the treatment and the absence of permanent cures. His supporters argued that centers contained a multidisciplinary team who all contributed to the care of the patient and his family. But some challenged the utility of the center by suggesting that that the centers geographically isolated cancer patients and often divided the family between home and the institution. One participant used the term "fatherectomy" to describe the separation between the ill child and the mother from the father.[49] Another participant countered that parents felt more secure when their child was treated in a center with other children with leukemia or cancer, saying, "Many, many parents have said to me, 'I am glad I came to this center because here I feel the security of all the doctors interested in the field.'"[50]

Addressing these concerns, however, was futile if local pediatricians did not refer their patients to centers. Without a standardized system, care was provided in a variety of settings using different approaches. Charles J. A. Schulte of the U.S. Public Health Service said, "Less than a third of children who have leukemia are cared for by large institutions devoted to the care of patients with malignant diseases or by physicians involved in large cooperative studies."[51] The discussion regarding the proper relationship between the pediatrician and the cancer specialist exposed the emerging tensions between the two groups of physicians. One participant presented the views of a general pediatrician as, "How can I get rid of this patient? Nobody likes doctors who take care of dying kids; how can I unload this patient?"[52] Another reported that he had observed an attending physician bypass leukemia patients during rounds, explaining, "Well, it's a hematology patient."[53] Hartmann confirmed, "No one wants to take care of dying children by preference. Most of us have been hematologists (specialists in blood disorders) and have gotten into the field of oncology (cancer specialists) because of the therapeutic effects of chemotherapeutic agents, especially in children with leukemia."[54]

In short, Hartmann saw the new set of specialists as having a more positive, active orientation toward the treatment of this set of diseases.

The 1960s were a decade of transition for the patient with cancer and for the professional at the bedside. Many critics of chemotherapy remained. Surgeons and radiotherapists defended the traditional cancer treatments that they could offer, while others actively criticized the limited efficacy and harsh side effects of chemical therapies. In this setting, hematologists and a nascent group, medical oncologists, vied for control over this new treatment modality.[55] When they learned that chemical agents had induced temporary remissions in the leukemias and lymphomas, some hematologists began to refocus their work from problems of classical hematology—like the anemias and coagulations—to cancers of the blood and related tissues. Simultaneously, a small group of internists became interested in the application of chemical therapies to common solid tumors, like those of the breast, but when demonstrable gains in the treatment of solid tumors proved to be limited, they extended their reach to include all cancers. In their early organizational meetings, those promoting medical oncology defined the burgeoning medical subspecialty as devoted to the "total management" of all patients with cancer. They accepted full responsibility over coordinating surgical or radiation treatment, administering chemotherapy, rendering supportive care if complications occurred, and tracking patients from their initial diagnosis through cure or end-of-life care. The two professional groups staked a claim over the same patient population: the "unwanted patient" was now a contested patient.

Through a series of heated internal discussions and cross-disciplinary debates, each group defined its identity, training requirements, and targeted patients. Like cardiologists, both groups pursued (and gained) subspecialty status by building alliances with the American Board of Internal Medicine. It was hematologists, though, who perceived that they were gradually losing status to the emerging group. Medical oncologists (first unified under the banner of the American Society of Clinical Oncology in 1964) then compounded these concerns by initially proposing a plan that ensured complete separation between the two groups. After further consideration, they approved a compromise that recognized the value of joint training and certification in hematology-oncology. This specialist was uniquely prepared to cope with the challenges posed by a child diagnosed with leukemia now that multiple agents could be offered.

While the involvement of a specialist was key, the "best" case was described as a cooperative partnership that formed when the primary physician continued treating the child with cancer under a center's direction. Denman Hammond, the head of the Division of Hematology at the Children's Hospital of Los Angeles, said that all of his patients were referred to the center, so there was always a primary physician involved in the case (although sometimes at great distance). He stated,

> We, in talking to the parents, make it very clear to them in the first interview that we are happy to try to give them complete care for leukemia but that the family physician is very important to them. We would hope that the child would be in remission for 80 or 90% of the course of leukemia. During this time, the child is going to have colds and infections with other pediatric problems they may have otherwise; and, appropriately, the family physician is the one who should take care of these problems.[56]

In this case, hematologists or oncologists acted as consultants who saw well patients every month or two and sick children during relapse or terminal care. A lack of regular communication and inadequate education were cited as two obstacles to maintaining a good working relationship between the primary physician and the specialist. One participant complained that there was an attitude conveyed by specialists that community physicians did not know enough about this "field of expertise" and that they should not be treating the patients. Hartmann suggested that by implementing an elective two- or three-month rotation for second-year pediatric residents to introduce them to protocol studies and available drugs, it would enable physicians at cancer centers to work with experienced physicians in outlying communities (though not always the referring doctor).[57]

Participants presented two studies, based on the Children's Hospital of Los Angeles and the Pacific Northwest Children's Cancer Center in Seattle, as evidence of the advantages conferred in specialized cancer centers. Data collected from the Children's Hospital of Los Angeles documented shifts in the location of patient care, specifically in the proportion of time that a leukemic child spent in the hospital during the course of his or her disease. "Prior to 1953, when we had no cooperative programs nor much in the way of chemotherapy," Hammond recalled, "the course of leukemia in children was rather short. The records indicated that about 30% of the course, from diagnosis to death, was spent as an inpatient in a hospital."[58] In the next period, 1953 to 1957, patients received more chemotherapy and transfusions. Median survival time increased to about a year, but the children spent only about 13 percent of their illness in the hospital. In the most recent period, 1957 to 1963, median survival improved to about a year and a half, but the period of hospitalization dropped to 4.1 percent.[59] As the duration of the disease lengthened, there was a shift from treatment in the inpatient wards for the duration of the disease to ambulatory care during acute exacerbations.

The Pacific Northwest Children's Cancer Center, a referral service for children with malignant diseases from the Seattle metropolitan area and Alaska, also established outpatient clinics and many affiliated services to provide care. The hematology and oncology clinic had grown from one small room to a four-room outpatient clinic with four physicians to manage the intravenous medication, bone marrows, and blood transfusions previously provided by the hospital inpatient service.[60] Medical care was provided through a weekly tumor clinic, a daily hematology-oncology clinic, and a 24-hour a day on-call physician. An ambulatory transfusion clinic was conducted each morning, afternoon, and evening for children to receive transfusions, plasma, or platelets in an effort to keep children out of the hospital as much as possible. Children who needed to be admitted to the hospital for longer than six to eight hours to evaluate signs of toxicity or relapse and make major changes in therapy were admitted on a one-day service. However, the charges were reduced to a half to a third of the usual inpatient charge through a special arrangement with an insurance carrier. The National Cancer Institute also provided limited travel funds and support for children using outpatient services as part of cooperative protocol studies.

Finally, participants debated whether enrollment in cooperative studies constituted the best treatment for every child. Some offered that cooperative studies added to the knowledge of the disease through research and potentially contributed to the welfare of other leukemic children. They also provided solace for parents, since they could now view their child's life as ultimately purposeful, as helping another through their suffering. Despite these efforts, concerns arose regarding the conduct of the cooperative studies and their risk to the individual patient. At the National Cancer Institute, one major investigator received repeated criticism from his colleagues for not following proper FDA animal and toxicity trials before testing the experimental regimens on children.[61] Possible abuse was a major concern from inside and outside the medical profession. Also, the VAMP-type protocols required National Cancer Institute researchers to readmit children in complete remission to the hospital for additional experimental chemotherapeutic treatment cycles. These additional cycles—given to children who felt healthy—made them feel extremely ill. While parental permission was required before submitting children to these risks, few had refused the experimental treatments at National Institutes of Health.[62] At the 1966 conference, Hartmann challenged his colleagues, "Is it morally and ethically right to treat a child who is in a terminal state of his disease with these compounds?"[63] He urged his listeners to weigh several factors when making their decision: the short remission gained by the patient, the patient's quality of life, and the benefits for other young sufferers. He insisted, "We must often and always reflect on whether or not this is the right thing to do. There is no other way to find out whether such agents are effective, but we must also look upon the child and the family group as individuals."[64] In the 1960s, the availability of increasingly complex, aggressive chemotherapy at a limited number of cancer centers and the life-threatening side effects that often accompanied combination treatment required physicians to carefully consider a calculus that weighed their scientific goals against those of each child and family coping with cancer and, at times, to rigorously defend their decisions.

Death and Dying

In *Blood of the Lamb,* Wanderhope attempted to preserve many of the family's normal daily routines while accommodating the constant demands of his daughter's illness and treatment. Carol and her father continued to read, listen to music, and drink hot chocolate late in the evening before bed, but there were also perceptible changes around the house. Silence stifled any discussion of Carol's illness. At the time of Carol's diagnosis, Scoville had implied that a measure of privacy—perhaps even secrecy—about the disease was desirable, suggesting the name of a tutor who could aid Carol in her schoolwork who "won't talk" or "even ask any questions."[65] Later, he recommended that Wanderhope tell Carol that she had anemia because, "That's part of it, after all."[66] Wanderhope and the family's housekeeper, Mrs. Brodhag, were careful to hide the truth about the illness from the eleven-year-old girl, but he found it difficult not to spoil his daughter.[67] He showered Carol with gifts on her twelfth birthday, buying her a bicycle, half a dozen dresses, shoes, ballet leotards, books, jewelry, a doll, and an expensive tape recorder. Brodhag admonished him to limit the number of presents he purchased and to be less obvious about recording her piano playing. One evening, Wanderhope found Carol watching a documentary on cancer on television. The program showed Carol's physician, Dr. Scoville, examining a young boy in the pediatric cancer clinic. The narrator announced, "The most fruitful source of study, and the best variation of the disease in which to try out certain new remedies, is that form in which it cruises in the bloodstreams of children under the name . . ."[68] Her father quickly diverted her attention away from the television program. Despite mounting evidence that suggested otherwise, Wanderhope and Brodhag tried to adopt the approach of the Children's Pavilion at home—to maintain the attitude that "Everything Was Fine."

Medical innovations had created a young patient population that included more children with chronic, prolonged, and fatal illnesses; however, only a few studies systematically investigated truth telling or death and dying in children before the late 1950s and 1960s. Truth telling to adult cancer patients was also under intense scrutiny at this time. In 1962, Lemuel Bowden, a physician at Memorial Hospital, editorialized, "It is absurd for a physician who has spent many years studying the science and practicing the art of medicine to discuss the diagnosis, methods of treatment, and the probable results of therapy in detail with a patient who, at best, can bring only rudimentary comprehension to the problem at hand."[69] For Bowden, news about incurable cancers should not be delivered to the patient, but sweeping changes in accepted practice and policy were already underway. A study published a decade later revealed that whereas 90 percent of doctors had not discussed diagnoses with patients in 1962, 90 percent reported that they now had adopted a frank approach.[70]

A similar debate played out in medical and psychosocial journals over truth telling to children of all ages and stages of development. In the 1950s, child psychologists Morris Green and Albert Solnit advocated open communication between the child, physician, and parent but did not favor truth telling in all cases.[71] Green cautioned physicians to carefully consider a child's age and developmental stage and noted, "A concept of death does not become established in most children until just before puberty or early in puberty."[72] Based on his own clinical experience, he found that children with fatal illness did not directly ask if they were going to die, but inevitably sensed what was happening to them or in their family. Green advocated an individualized approach to truth telling but also recommended that children's queries never be completely evaded.

In 1965, Joel Vernick, supervisor of the social work department at the National Cancer Institute Clinical Center, and Myron Karon, chief pediatrician of the medical branch of the National Cancer Institute, added their analysis to the growing body of medical literature. Their article about telling leukemic children the truth about their disease—"Who's Afraid of Death on a Leukemia Ward?"—departed significantly from a number of previous authors who had recommended that diagnoses be kept from the child to protect them from unneeded anxiety.[73] Vernick and Karon gathered information from fifty-one children from nine to twenty years old who were hospitalized for acute leukemia at the National Cancer Institute. They collected data using "life space" interviews that enabled workers to focus on the actual events that concerned the child, such as an intravenous infusion or a pill.[74] They also organized weekly group sessions so that parents had a forum in which they could discuss their concerns with the senior staff—members of the staff who formulated long-term research goals, but were also interested in each individual child's case. Vernick and Karon found that these meetings usually began with medical matters but moved to emotional topics.

The authors concluded that the central question was "not whether to talk to the child about his serious concerns, but how to talk to him."[75] Lies temporarily deceived the child, but they also raised a child's anxiety and caused serious behavioral problems. The result was often irreparable damage to the parent-child relationship. In addition, the authors discovered that the sudden absence of the child and his or her name plate outside the room, the removal of equipment including the oxygen tent, the disconnection of the respirator, or the transfer of a child to a separate room in the middle of the night served as clear nonverbal cues to children that a death had occurred on the floor. Straightforward yet age-appropriate discussions with the children were needed.

The medical staff at the National Cancer Institute changed its policies in order to address issues of truth telling and death. Following a death, the first staff member to see other children discussed the event directly. To comfort other children, the staff member emphasized that the child who died was very ill (unlike others on the floor). "With the cooperation of the parents," Vernick and Karon wrote, "every child over the age of 9, and in some instances those even younger were told the diagnosis of their illness as soon as the presence of the disease had been verified."[76] Older patients, they suggested, should be informed about procedures or the course of treatment not for legal consent but to gain cooperation and understanding. Although the amount of shared information varied depending on the child's age, the authors emphasized open communication between physicians and other allied health workers on the cancer floor, parents, patients, and their siblings to abolish secrecy and invite dialogue about disease and death.[77] Karon and Vernick concluded that the care of the fatally ill child was perceived as one of the most difficult tasks faced by physicians, because they were not properly trained for it in a medical curriculum focused on curing illness, not preparing for death and dying.

In September 1965, the editors of the *American Journal of the Diseases of Children* solicited a response to the Vernick and Karon article from Joseph H. Agranoff and Alvin M. Mauer, physicians from the Children's Hospital in Cincinnati, Ohio, who wrote, "only discussion and controversy can ultimately lead to progress."[78] Contesting the conclusion that older children with leukemia could and should be told the truth, the Cincinnati doctors suggested that the children in the original study were a special population not representative of all pediatric cancer patients. They pointed out that the children treated at the National Cancer Institute were hospitalized for relatively long periods of time on a special study ward reserved for children with acute leukemia. Children attended school in separate facilities provided by the National Institutes of Health and, when it was deemed medically permissible, children participated in weekly activities and outings in the Washington, D.C., area like swimming, trips to the zoo or museums, and White House tours.[79] Under these circumstances, children closely identified with other dying children, which made it nearly impossible to conceal the nature of acute leukemia.

Unlike the National Cancer Institute, children at the Cincinnati Children's Hospital were admitted to a general medical ward that contained patients with a variety of illness. Children with leukemia spent as much time as possible at home, were only admitted to the hospital during an acute phase of their illness, and made return visits to a hematology clinic—not a special leukemia or cancer clinic where they would only be exposed to other children who shared their diagnosis. Thus, deaths on the general ward were uncommon and children did not receive the same exposure to illness and death. Few children asked about their illnesses. They worried that applying Vernick and Karon's recommendations to children treated outside of the National Cancer Institute environment would increase anxiety and encourage families to rely on information about acute leukemia from sources outside the hospital. Agranoff and Mauer argued that their system at the Children's Hospital represented the prevailing care for children with acute leukemia and that this model dictated different policies. While agreeing that Vernick and Karon persuasively argued their position, the authors called for additional information gathering so that informed decisions could be made about the management of the child with leukemia cared for in the general medical center.[80]

In their rebuttal to the Cincinnati doctors' comments, Vernick and Karon noted that similar concerns had hindered investigations about truth telling in the past. They acknowledged that their ward was not typical but claimed children's anxieties and perception of their environment were universal. It might be even more important to openly share information in a general ward, they argued, because such an environment deprived children of the opportunity to communicate easily with others suffering from the same disease.[81]

Despite this counterargument, Agranoff and Mauer's views resonated with other physicians. Henry F. Lee, a physician in the Chestnut Hill Pediatric Group in Philadelphia, wrote that his experience in a children's medical center and in the pediatric service of several general hospitals taught him that older children, except, perhaps, those treated in a specialized leukemia ward, should not be told the name or nature of their diagnosis. Lee wrote that most of his leukemia patients had been satisfied with his false or partial diagnoses. He told patients that they suffered from "an unusual form of arthritis" or a "big spleen," so they would be spared frightening articles or conversations about acute leukemia.[82] In his own practice, Lee preferred to keep the child at home as much as possible and to minimize medical interventions, saying, "I am certain that my recent leukemia patients are far *happier* than those I cared for earlier when driven by more need within myself to achieve a 'university-hospital-type' workup and therapy."[83] He nevertheless acknowledged that his community hospital approach was ill suited to making advances in the disease. Like the Cincinnati physicians, Lee agreed that the approach described by Vernick and Karon should not be applied to all settings. Alfred Hamady, a physician from Battle Creek, Michigan, agreed with many of Lee's arguments. He recognized the value of truth telling but gave precedence to ensuring the quality of the child's final days. He also expressed skepticism that the methods were widely applicable. "Most of the advice [about what to tell child with leukemia] is coming from 'large' centers in the form of generalizations based on 'controlled studies,' 'statistics,' and the like. How can the subject of impending death, as it relates to a child, be anything but a personal and intimate matter? How can controlled studies and statistics take into account the close, hour-to-hour relationship between the young patient, his playmate, his parents, and the attending physician?"[84] Instead, Hamady preferred that no standard practice for truth telling be established but, rather, that each case should be considered individually based on information exchanged through late-night phone calls from parents to physicians, from the child's teacher, home visits, and other conversations. He maintained that preserving the positive aspects of the child's remaining life was at the foundation of his recommendations.[85]

The exchange printed in the pages of the *American Journal of the Diseases of Children* again demonstrated the vast ideological differences that existed among National Cancer Institute physician-researchers, physicians at specialized children's hospitals, and community doctors. Representatives from each group argued that their disparate treatment methods, aims, and outcomes necessitated a different truth-telling policy from the one recommended by Vernick and Karon and pitted the results of the National Cancer Institute study against personal experience, anecdotal evidence, and familiar practices. A single path did not prevail in the 1960s, but it was increasingly clear that parents had vital responsibilities and burdens as caretakers at all stages of treatment or disease.

Studies on the relationship between physicians, parents, and physicians generated standardized guidelines for cancer care in the 1960s.[86] In his research, Stanford Friedman, senior instructor in pediatrics and psychiatry at the University of Rochester Medical Center, charted modifications in parental behavior toward acute leukemia and the threat of death as the face of the disease changed from the 1940s through the 1960s.[87] Friedman found that most parents seemed to suspect the seriousness of diagnosis before hearing it from the physician and hypothesized, "This probably reflects the relative sophistication of the general population regarding medical matters, particularly about diseases that are associated with various fund drives."[88] Parents also combed newspapers, magazines, and medical texts for answers to their questions, especially about etiology. To remedy any misinformation or confusion from other sources, he urged physicians to distribute the National Cancer Institute publication "Childhood Leukemia: A Pamphlet for Parents" as a dependable source.[89] Pamphlets, however, could not possibly prepare parents for the challenges that arose.

By systematically studying behavior and physical indicators of stress, Friedman and his team monitored forty-six parents of children with fatal, neoplastic disease (primarily leukemia) while the children were being treated at the National Cancer Institute. They quantified parents' responses to events during their child's illness in order to determine the primary stressors. For example, as chemotherapeutic regimens began to reliably induce lengthy remissions—a long period when their children were apparently restored to health—parents remained in denial longer. A child's first medical relapse often represented the first time his parents were confronted with the reality of his disease. The sudden downturn provided a powerful visible reminder that the child suffered from a critical illness, and parents commonly experienced a major crisis at the time of this event. The longer life produced by chemotherapeutic agents also lessened parents' grief at the time of death by giving them more time for anticipatory grief. However, Friedman advised physicians to contact parents in the months after a child's death to provide an opportunity to discuss the illness and death again.

Friedman also drew more general conclusions. Weekly interviews, ward observations by the psychiatrists and nurses, and data from a regular parental discussion group revealed that the parents—classified as mostly lower and middle income, predominantly white and Protestant, and from both urban and rural locations—maintained relationships, preserved their own personalities, and carried out necessary tasks during their child's period of illness. Significantly, the authors detected an overall pattern of coping strategies, defenses, and searches for meaning in their child's illness and death that illustrated a common "natural history" or sequence of adaptational techniques employed by the group that allowed hospital staff to implement additional services like support groups or individual counseling for the parents of children cared for in leukemia wards or clinics.

Studies by Friedman and others suggested certain behavioral patterns among parents of leukemic children at the National Cancer Institute, but characters in *The Blood of the Lamb* illustrated the myriad variations possible in parents' approach to the disease. De Vries created a vocal mother in his novel, an obese woman clad in a rumpled housedress who chain-smoked cigarettes as she paced the halls of the children's cancer ward. A short, darkly comedic monologue suggested her lack of understanding about the disease:

> "Boy, dis place," she said. "When me and my little girl come in here, she di'n't have nuttin' but leukemia. Now she's got ammonia." I listened, unbelieving. "Ammonia. Dat's serious. She's in a oxygen tent, and I can't smoke there. It's a tough break for her because, like I say, at first she di'n't have nuttin' but a touch of leukemia. I don't believe I ever hear of dat before. What is it?"[90]

The woman's daughter died of pneumonia soon afterwards. Other parents in the book clearly understood the stages of acute leukemia and became grief-stricken as their children's lives neared their ends. Wanderhope watched one father's frustration escalate into a physical struggle with a cancer specialist after he received the news that all of the therapeutic options had been exhausted in his child's care.[91] A third parent, a mother who had seemed outwardly staid to the other parents visiting the ward, unexpectedly confided her worries to Wanderhope in the middle of the night as they rested near their children's bedsides. Mistaking Wanderhope's late-night laughter for stifled cries of despair, she divulged that the demands of her daughter's illness had rendered communication with her husband nearly impossible. She welcomed the opportunity to talk to another parent on the ward who could fully understand and relate to her concerns.[92]

The reader could gain insight into Wanderhope's personal experience of his daughter's illness through two intertwined themes—his depiction of Scoville and medical science, more generally, and his inner debates about religion. From Carol's diagnosis to her death, Wanderhope viewed medicine with both respect and skepticism. At times, Wanderhope gained optimism from Scoville's inexorable scientific curiosity, his single-minded dedication to his profession, and his aggressive approach toward the medical management of acute leukemia. At the end of Carol's first hospital stay, De Vries depicted Scoville as virtually superhuman. Wanderhope noted,

> When I saw him enter the ward on the third afternoon he looked sixty years old, rumpled and unshaven, scarcely able to hold the dispatch case he was dragging in one hand. He had flown to five cities in a series of research conferences, with a dash to Washington to bludgeon loose some funds for an experimental drug costing fifteen thousand dollars a pound, and he had not slept in a bed, he told us cheerfully at the foot of Carol's, for thirty-six hours.[93]

But Scoville's positive outlook was tempered by cynicism from parents who had already seen their child alternate between remission and relapse. In the early stages of Carol's illness, as Scoville lauded the chemotherapeutic agents developed during the past ten years, one father noted, "So death by leukemia is now a local instead of an express. Same run, only a few more stops. But that's medicine, the art of prolonging disease."[94] As his daughter's condition worsened, Wanderhope adopted some of this rhetoric. He likened the doctors' white coats to "butcher's coats" and said that they were worn as they "merely hounded the culprit from organ to organ till nothing remained over which to practice their art: the art of prolonging sickness."[95] As Carol alternated between relapse and remission, sickness and health, the hospital and home, Wanderhope's hope only wavered intermittently, but his faith in a benevolent God and the miraculous power of St. Jude, the saint of hopeless causes, completely faltered.

In *The Blood of the Lamb*, a final burst of rigorous, experimental treatment marked the final phase of Carol's cancer therapy. After Carol became resistant to methotrexate, Scoville prescribed a second course of steroids. Because steroids were less effective when administered the second time, he supplemented them with an experimental drug that had induced shown short-term activity in preliminary cooperative clinical trial results. Before her next marrow count, the drugs had begun to successfully kill Carol's most quickly replicating cells. Thus, she began losing her hair (alopecia) and contracted an infection when her healthy white cell count fell. She received broad-spectrum antibiotics and was readmitted to the hospital as a precaution, but, fortunately, a bone marrow test showed that she had again achieved a full remission. On his way to the hospital to celebrate the remission with Carol, Wanderhope briefly stopped into a church near the hospital, a routine that he had developed during his daughter's illness. In the pews of the church, another parent informed him that an infection had swept the ward and that half of the children had been placed in oxygen tents and on antibiotics to guard them against the threat. Wanderhope rushed to the children's ward with hope that his daughter had been spared from the outbreak. Tragically, the infection had already ravaged Carol's vulnerable system and covered her lower body in discolorations from the blood poisoning. Unable to treat the rampant infection, Scoville advised the grieving father to allow him to halt all life-saving measures and use morphine to lessen the child's pain. Carol died within hours.

The Blood of the Lamb reflected the experiences of Emily and Peter De Vries through the lenses of Carol and Don Wanderhope. Although the novel was a fictionalized account of the DeVries's experiences, the two well-developed characters provided valuable insight into patient and family experiences during a critical decade of incremental developments in acute leukemia research and therapy. This was especially valuable for a period in which few parents spoke publicly about their experiences in an extended form. The novel and the conference proceedings from "Care of the Child with Cancer" offered two complementary sources for analyzing changes in medicine that contributed to a growing distance and distrust between the physician and the patient in the 1960s and 1970s. The Wanderhopes received Carol's diagnosis at a home visit from their family physician, a rarity in the 1960s. After this point, she was regularly treated in a specialized hospital away from her home. Wanderhope initially respected Scoville's expertise and welcomed the inclusion of experimental therapies in his daughter's care. During Carol's last crisis, though, Wanderhope questioned the promises of medical science and his religious faith. He referred to leukemia as the "Slaughter of the Innocents" and asked, "Who creates a perfect blossom to crush it? Children dying in this building, mice in the next."[96] Parents labored under the struggles presented by acute leukemia—its devastating effect on their child's health, its ability to erode hope at a time of incomplete cures, its threat to the stability of marriage and family, and its fundamental challenges to their beliefs and values.

At the time of Carol's death, chemotherapeutic agents prolonged children's lives, but "cures" of acute leukemia were viewed as miracles. In a well-publicized example, *Life* magazine published a two-page story on Mother Elizabeth Ann Bayley Seton, the first native-born American likely to achieve sainthood.[97] After Seton's conversion to Catholicism in 1805, she founded the Sisters of Charity, America's first society of nuns, and aided in the establishment of the United States parochial school system. In 1963, two American cardinals, members of the Daughters of Charity, and thousands of other religious pilgrims traveled to St. Peter's Basilica in Rome for a ceremony that recognized two cancer cures "as the direct and miraculous result of prayers to her."[98] One of the cured cancer sufferers was fifteen-year-old Ann O'Neil, a leukemia survivor who traveled to Rome from Baltimore, Maryland, to observe the beatification proceedings. A study published in 1951, one year before O'Neil's diagnosis, reported that only three of 150 child and adult patients with acute leukemia at Memorial survived more than a year, and all died within fourteen months.[99]

By the mid-1960s, the few acute leukemia survivors were no longer only those healed by "spontaneous cures" or religious miracles. In 1964, Burchenal was able to present dozens of cases of long-term acute leukemia survivors (children and adults) who had been off therapy for at least a year and were living with no evidence of disease.[100] Burchenal gathered these cases by surveying hematologists from around the world and recording the data in the Acute Leukemia Long-term Survival Registry of the Acute Leukemia Task Force. Although he estimated that they represented only between 0.1 and 1 percent of all cases, the statistics provided solid evidence that long-term survival was possible by combining chemotherapeutic agents and providing adequate supportive therapy.[101] Burchenal boldly predicted that these results suggested the eventual control of acute leukemia, a childhood killer. News of long-term survival in acute leukemia supported the proposed goals of the "War on Cancer"—a large, federally funded program against cancer declared in 1971 that targeted the rapid elucidation of effective therapies and cures for the dread disease.

CHAPTER SIX

"The Truly Cured Child"

PROLONGED SURVIVAL AND THE LATE EFFECTS OF CANCER

During the 1970s, aggressive chemotherapeutic therapy led to a dramatic improvement in the long-term survival rates of children with leukemia and other common childhood cancers. Despite its name, acute lymphocytic leukemia, the most common type of childhood cancer, resembled a chronic illness for many young victims by the end of the decade and years to come. An essay by Amy Louise Timmons, an eleven-year-old girl with the disease, published in the *Journal of Pediatrics* and the *American Journal of Nursing* was one girl's plea for parents and health professionals to consider the period of healthy life enjoyed by many young patients.[1] In May 1974, Timmons had already lived nearly three years beyond her physician's initial prognosis and had remained an honor student, the president of her class, and a member of the Girl Scouts. She beseeched readers not to pity her; she had led an active, satisfying life despite her illness and believed that all children should be allowed to lead near-normal lives when they felt well. Timmons referred to an episode of the popular television program *The Waltons* in which a mother restricted her teenage son to his room after his leukemia diagnosis. Following the boy's appeals to be let out of the protection of their home, his mother finally relented and he rejoined his friends outside. Timmons acknowledged that treatment for leukemia caused children to experience infections, hair loss, nausea, and soreness at injection sites but informed readers that children often felt well enough to enjoy life. Only a month after composing this essay, Timmons spent a long day swimming at the beach and playing games with her family until very late in the evening. The next day, she entered a coma and died before she could be readmitted to the hospital. Combinations of chemotherapeutic agents and supportive therapy in leukemic children allowed Timmons and other children to live years after their initial diagnosis. And, building upon Burchenal's earlier claims, physicians tentatively suggested that permanent cures had been achieved in this set of patients.

In the 1970s, the history of pediatric cancers splintered into three separate narratives: illness and incurability, short- and long-term survival, and death and dying. Progress in treating and curing childhood cancers became the justification for increased cancer research funding, since gains in the five-year survival rates in many childhood cancers were often upheld as a predictor of future results in adult malignancies; however, the transformation of leukemia and other cancers from acute, invariably fatal diseases to nearly chronic conditions caused unforeseen challenges for physicians, allied health professionals, parents, and children. Surgery, radiation, and chemotherapy not only cured but also produced debilitating side effects such as recurring tumors and physical, mental, and learning disabilities. As child survivors lived longer, they faced discrimination among their friends, at school, by insurance companies, and in the workplace. For some, the manifestations of cancer often lasted long after childhood. For others, the late stages of their illness demanded a reconsideration and renegotiation of a "good" death.

Illness and Therapy

In the 1970s, longer survival for children with acute leukemia presented new challenges for medical professionals responsible for designing and administering combination chemotherapy regimens and managing the patient throughout all stages of disease. Chemotherapy protocols induced complete remissions in 90 percent of children with acute lymphocytic leukemia.[2] Even after a complete remission had been achieved, the child remained on chemotherapy from two to five years in an attempt to completely eradicate the leukemic cells and restore normal function.[3] Unfortunately, many children who experienced these prolonged periods of treatment eventually succumbed to a recurrence of leukemia in the central nervous system. Beginning in the 1970s, physicians ordered rigorous treatment including irradiation of the cranium and spinal axis, intrathecal administration of medication, or both to prevent the development of life-threatening disease in this physiological sanctuary.[4]

Many other medical interventions were required to maintain the health of leukemia patients in this tenuous state. Intensive antibiotic therapy was used during periods of immune system suppression. Despite these prophylactic efforts, 60 to 80 percent of deaths in children with leukemia treated at National Cancer Institute between 1965 and 1971 were caused by viral, fungal, or bacterial infections.[5] Chemotherapeutic drugs attacked both proliferating leukemia cells and rapidly reproducing normal cells indeterminately, causing hair loss and ulceration of mucous membranes of the gastrointestinal tract from the mouth to the rectum. Transfusions of packed red blood cells helped children with anemia; transfusions of concentrated platelets aided those bleeding from inadequate production of these clotting tools. Frequent hospitalizations and supportive measures were required to stabilize young patients.

Despite these setbacks, treatment results compiled from the seventeen centers involved in research and treatment programs dedicated to childhood cancer centers reported a 50 percent five-year survival for children suffering from acute lymphocytic leukemia, a greater than 50 percent five-year survival for Wilm's tumor and soft tissue sarcomas (tumors in muscle, fat, or blood vessels), and an 89 percent five-year survival for Hodgkin's disease (cancer of the lymphatic system, part of the immune system).[6] Osteosarcoma and Ewing's sarcoma (bone cancer commonly found in the pelvis, thigh, or shin) also had rates near 50 percent but not at the five-year mark. Brain tumors and neuroblastoma were the only common childhood tumors with disease-free survival rates significantly lower than 50 percent.[7]

A single published source could not fully capture children's experiences with cancer, particularly because only about half of children had access to major cancer treatment centers or National Cancer Institute–related treatment research groups.[8] The following year, an American Cancer Society publication documented quantitative data based on participants in the End Results Group, a national program sponsored in part by the National Cancer Institute to evaluate the efficacy of cancer therapies.[9] In this report, data was submitted by more than 100 hospitals of varying types and sizes from different regions of the country. Thus, small community hospitals, larger general hospitals, and facilities affiliated with medical schools shared their data to provide a broader cross-section of children's outcomes. Based on this broader sample, only 5 percent of children with leukemia lived five years after diagnosis.[10]

What factors led to the disparities in survival rates reported in the two sources? Distinct patient populations were a major factor. Joseph Simone, a physician at St. Jude Children's Research Hospital in Memphis, Tennessee, a new center for the research of childhood cancers and other catastrophic diseases, described the admissions:

> Most of the children were from the mid- or Deep South, the area of the United States with the lowest per capita income, lowest hospitalization insurance cov-

> erage, the highest infant morality rate, and the shortest median life span. Since this hospital is the only pediatric medical center in the area that is completely free and open to all social, economic, and racial groups, it tends to attract low-income families and thus, many children with neglected health, advanced disease, and poor nutrition. All eligible children referred here are registered in the studies and included in the reports, regardless of their condition on admission. This includes children who receive little or no therapy, who die of overwhelming leukemia, infection, or hemorrhage shortly after admission.[11]

The particular mission and research guidelines that shaped hospitals like St. Jude attracted the sickest children.

Two inseparable, but sometimes contradictory, purposes for young patients—cancer trials and access to advanced care—continued to be touted as advantages of referrals to specialized institutions. In 1978, at the American Cancer Society–sponsored National Conference on the Care of the Child with Cancer, Giulio J. D'Angio, director of cancer centers at Philadelphia's Children's Hospital and the University of Pennsylvania, discussed cancer centers as sites of research collaborations between trained medical specialists in the field. He declared, "Highly coordinated battle plans now are drawn up by integrated staffs of surgeons, radiation therapists, and chemotherapists. Each move is plotted in advance and carried out with military precision."[12] They then pooled their results with other institutions involved in the national cooperative groups.

D'Angio and other specialists tirelessly promoted their principles and practices in order to gain wider support for their nascent field. In an opinion piece about the best medical care for the "hopeless" patient, Emil J. Freireich, chief of research hematology at M.D. Anderson Cancer Center, wrote, "I don't mean to put down the man in private practice," but described how clinical medicine differed. Adopting the rigorous language attributed to basic science, he characterized clinical medicine as objective and the results as reproducible. Clinical researchers not only prescribed treatment, they employed controls and analyzed outcomes data to apply the knowledge gained from one patient to another with the same disease. He stated, "As a result of clinical research, any doctor now can give leukemic children a two-drug combination—vincristine and prednisone—and produce remissions in about 85 percent of the cases. But every doctor shouldn't—he won't know what to do for an encore."[13] A single remission was only the first and most basic step of leukemia treatment. Freireich warned readers that only professionals in clinical research centers had access to information about new discoveries from cooperative studies before they were published and could rapidly adjust the dosage or the method of drug administration if subsequent complications occurred.[14]

Freireich explicitly tried to dispel concerns that patients were treated as "guinea pigs," but his arguments suggested otherwise. He insisted that patients made the conscious choice to participate but acknowledged that drugs often made patients sicker than before. He mocked those who advised, "Let the patient die with dignity" and urged physicians to fight as long as the patient remained alive. Freireich said, "You give up when you can't get any blood into a vein. As long as she's breathing and her heart's beating, tend to her. You can't know what will happen."[15]

As pediatric cancer treatment became part of a culture of clinical experimentation in the United States in the 1950s and 1960s, the roles of physician, patient, and hospital ward had become intertwined with those of investigator, research subject, and laboratory.[16] Multiple factors contributed to this culture of investigation: the discovery and development of "miracle drugs" that used acute leukemia as a model; the increased role of cancer charities including child-focused organizations like the Jimmy Fund and the Leukemia Society of America; an influx of public funds into cancer research; links between government-sponsored research and industry; and a concentration of patients in specialized institutions such as comprehensive cancer centers.[17] Parents weighed the potential benefits of experimental cancer therapy to their children, but they also considered how to preserve quality of life and limit suffering when giving proxy consent.[18] The Declaration of Helsinki's *Recommendations Guiding Doctors in Clinical Research* (1964) and the American Medical Association's *Ethical Guidelines for Clinical Investigation* (1966) distinguished between clinical research for patient care and nontherapeutic clinical research designed to accumulate scientific knowledge, but gave parents the power to consent for both types of research on children. In 1970, in *Patient as Person,* Paul Ramsey departed from these earlier recommendations and advocated that children not serve as subjects of medical experimentation unless "other remedies having failed to relieve their grave illness, it is reasonable to believe that the administration of a drug as yet untested or insufficiently tested on human beings the performance of an untried operation, may further *the patient's own recovery.*"[19] According to Ramsey, the health of the child, not the advancement of medical knowledge or the advantages for future patients, made medical experimentation on children permissible.[20]

The bold warnings of physicians at a 1977 symposium on the social and ethical issues in cancer prevention and therapy indicate possible conflicts of interest encountered by physicians. One participant complained,

> By declaring a war against cancer, and amplifying such jargon, and by providing strong financial persuasion for investigators to conduct experiments on human cancer patients today for the sake of the patient of tomorrow, an operative climate has been created that erodes . . . physician-patient relationships.[21]

By combining therapy and investigation, the loyalty of the physician was divided between the patient, profession, and national research goals. Phase I, II, and III human cancer studies mandated by the National Cancer Institute required different types of deliberation. Phase I studies evaluated a drug's toxicity in order to establish a suitable dosage for an evaluation of the drug's therapeutic effect in a future patient. This type of study required frequent clinical and lab exams to monitor toxicity. Direct benefits for the patient were not the primary objective of the trial. Phase II and phase III clinical trial studies offered more therapeutic potential, yet toxicity risks remained.[22] Each phase of experimental cancer treatment held the possibility of short-term and lasting long-term side effects like permanent damage to vital organs, mental abnormalities, and cancer. While oncologists characterized the majority of side effects as "temporary, predictable, and manageable" if monitored carefully by skilled investigators, critics decried the risks posed by cancer experimentation and advocated for greater lay autonomy.[23]

In contrast to Ramsey's revisions and recommendations, the recollection of one mother demonstrated that parents' complex decision making process included the potential benefits for all children with leukemia and the preferences of their own child. She candidly confessed,

> There comes a time when there is a fine line between whether continued therapy is for the sake of research for future leukemics or whether it is truly best for the child at hand. Del and I felt much sooner than the physicians that the time had come, but it was more a feeling from the heart than from professional objectivity. Eric, unaware of the full circumstances and that the odds were against him no matter what, had wanted to go on with treatment for a longer period than we felt necessary. Maybe we should have been more open with him; that is something we will never know.[24]

The arguments from the social and ethical issues conference and from the mother of a child with cancer demonstrated that physicians had vested interests in promoting participation in clinical cancer research. Some parents, however, may have continued to give the professional, "objective" opinions given by doctors or the health of other children with leukemia priority over the emotional or intuitive sense of family members to protect their child from undue harm.

The aggressive treatment model was paired with "total care" programs at the National Cancer Institute–designated cancer centers organized in the 1970s. The centers originated from grant programs established by institute in the 1960s and 1970s to support multidisciplinary cancer research. As a result of the National Cancer Act of 1971, a cancer centers branch was established to fund develop new institutions, using Memorial Sloan-Kettering and M. D. Anderson as models. Two years later, a cancer center support grant defined two classes of cancer centers, including a "comprehensive" group that was charged with "conducting long-term, multidisciplinary cancer programs in biomedical research, clinical investigation, training, demonstration, and community oriented programs in detection, diagnosis, education, epidemiology, rehabilitation, and information exchange."[25] Like the program implemented by Farber at the Jimmy Fund Clinic in the 1950s these institutions also attempted to integrate the services of medical specialists with those provided by social workers, psychologists, nurses, and chaplains to confront both medical challenges and the "human side" of the cancer problem.[26] Although delivery of this complex model was—and is—difficult to execute for every family, it was singled out as an ideal for children and adults in the 1970s.

Treatment began with the affected child, but, in the words of one American Cancer Society writer, "The whole family actually becomes the patient."[27] Social workers assisted parents with the overwhelming financial burden of receiving extensive cancer care for their children at a facility located at a distance from their home. Duke University Medical Center joined the comprehensive cancer center network in 1972 and raised money to staff their new pediatric research unit with its own "total treatment team." Children from throughout the Southeast traveled to Duke, so transportation, lodging, and the financial hardships needed to be addressed for each family. The Duke medical system had a financial scale for determining each individual family's ability to pay, and outside sources assisted families with the costs. North Carolina's Crippled Children's Fund, state sources, community agencies, and local churches also assisted families in need. In addition, Duke hoped to build satellite clinics to help alleviate the rigors of travel.

At other cancer centers, Ronald McDonald Houses enabled parents to participate in their child's care by providing inexpensive lodging for families of hospitalized children. In 1973, when his three-year-old daughter Kimberly was stricken with leukemia, Philadelphia Eagles football player Fred Hill recognized the lack of affordable housing for parents of hospitalized children.[28] During his daughter's three years of treatment, Hill and his wife often joined other parents in sleeping on hospital chairs and substituting vending machine snacks for proper meals. Hill worked with members of the staff in the pediatric oncology unit at the Children's Hospital of Philadelphia and the fast-food operator that lent its name to the project to build a place that acted as a "home away from home." With proceeds from the restaurant's green St. Patrick's Day shake, the first McDonald House was constructed in Philadelphia. And, with continued support, the facility was able to adjust its rates based on household income, thus enabling parents remain near their child without incurring prohibitive hotel charges or transportation costs.

Longer disease-free survival for many childhood cancer patients required greater emphasis on transitioning young cancer patients from hospital to home. By being involved in each step of their child's cancer care, parents became proficient in skills they would need to repeat at home. At Stanford Children's Hospital in Palo Alto, California, nurse-parent teams incorporated parents into their children's daily care. The parents lived at a hotel on the hospital property free of charge in exchange for helping to feed their child or administer injections. A representative from the program said, "The parents are a vital part of our program. They have the expertise in dealing with their child. They know them better than we do. Sometimes the doctors are even willing to admit that mother knows best."[29] Stanford staff also included cancer education for patients in their program, teaching children how to read and interpret blood counts for themselves. Proclaiming a 50 percent success rate in referral cases, they boasted that their cooperative approach had improved mortality rates and buoyed optimism in patients and parents involved in the program. Nurses acted as communication link between parents and the treatment team as the child transitioned from hospital care to regular outpatient clinic visits. This need for continuous nursing led to the professionalization of the pediatric oncology nurse specialist by the mid-1970s.[30]

Death and Dying

In the 1970s, many families suffered under the strain of caring for and losing a child at a time when children were championed—in the courts and the media—as special, privileged members of the population that needed to be protected. As one researcher noted, "At this time in our history and consciousness, in this youth culture, children are not supposed to die."[31] In the *San Jose Mercury News,* an article titled "The Sad Wait at Ricky's House" told the story of five-year-old Ricky Pineda, a local boy in the final stages of acute leukemia. Over the course of his illness, Ricky had been hospitalized for spinal meningitis and survived four relapses, but doctors now predicted that the boy had only six months to live.[32]

Like many parents, Gloria and Richard Pineda bore the weight of their son's extended and ultimately fatal illness. A study at Stanford University Medical Center supervised by David M. Kaplan, director of clinical social work, had found that nearly half of thirty-nine families of leukemic children studied had experienced major social, marital, or psychiatric problems. In sum, eighteen couples had divorced or separated after the child's death, existing marital problems had been exacerbated in eleven couples, the surviving children experienced difficulty in fourteen families, drinking problems developed in fourteen families, and fourteen families experienced problems at work.[33] Based on these findings, Kaplan had called the average two- or three-year period of illness a time "a state of siege" when families were forced to come to terms with their child's impending death, meet unexpected medical expenses, and care for other children. Jordan R. Wilbur, head of the Department of Pediatric Oncology at Stanford's Children's Hospital, argued that the results no longer applied to cancer cases because the study had been conducted in a traditional pediatrics department, not the new comprehensive care program that had implemented the total care approach.[34] Wilbur confidently opined that Kaplan's findings would no longer apply.[35] Such organizational changes may have relieved some of the hardships endured by parents, but the demands and uncertainty of Ricky's illness strained the Pineda's marriage and prompted Richard to threaten divorce.

Ricky's case also reintroduced questions about the emotional health of young patients. During his illness, Ricky became curious about his impending death. He wondered aloud whether he would be buried, whether he could keep his favorite possessions, and whether his mother would accompany him to heaven, questions that revealed the boy's understanding and confusion about the meaning of death. His mother lamented, "It's so hard to tell your child about death, but children have a premonition."[36] Gloria chose her responses carefully, but it was important to her to remain truthful to her son. In the 1970s, physicians' preferences toward truth telling continued to vary widely—some doctors advocated silence, others urged disclosure, and a third group thought that parents should make the final decision.

Investigators carefully listened to children's words and observed their behavior in order to gain insight about their understanding of illness and death and, consequently, what recommendations about truth telling were appropriate. By studying children's stories, researchers attempted to quantify children's anxiety about death.[37] Parents frequently insisted that their children did not know that they were suffering from a fatal illness, yet the researchers discovered that many children did comprehend that they were going to die, even if they had not been told directly. Many children who did not talk about their upcoming death (unlike Ricky Pineda) knew, however, that death was inevitable and deliberately concealed this information from their parents and the medical staff.[38] Myra Bluebond-Langner, the author of an influential book-length study, argued that the children's behavior reflected their socialization about death in America—that death should not be openly discussed. Bluebond-Langner urged parents and physicians to break the silence about death and dying and to be prepared for queries throughout the course of the child's illness.

Literature on truth telling published from the 1950s to the 1970s might have influenced the move toward more disclosure, but physicians' behavior also responded to social pressures that forced medical practitioners toward respecting greater patient autonomy.[39] By framing the model patient as one that was both autonomous and informed, supporters of the patients' rights movement attempted to empower patients and to restore their voice to the medical encounter.[40] Several components of this patient- or consumer-centered movement had a direct impact on the care of children with cancer, including the regulation of medical experimentation, wider acceptance of alternative medicine, and the new attention to death that gave patients and parents greater self-determination to reject life-extending medical care during the terminal phase of illness if they wished.[41]

Medical professionals' death anxiety and avoidance strategies continued to hinder interactions between physicians, parents, the young patient, and his or her siblings. In the 1960s, research by social workers and psychologists had begun to uncover this problem and design new strategies and solutions. At the end of the decade, Elisabeth Kübler-Ross published *On Death and Dying*, a best-selling book that implored readers to carefully consider and value the perspective of the terminally ill patient. Kübler-Ross described the account as "a new and challenging opportunity to refocus on the patient as a human being, to include him in dialogues, to learn from him the strengths and weaknesses of our hospital management of the patient."[42] By telling the stories of her patients and printing their own words, she sought to elevate the human above the medical technology surrounding modern death.[43] She criticized medical education that prioritized research and laboratory work over doctor-patient relations and recommended training for medical students that included psychosocial dynamics in the practice of medicine and curriculum on the care of the dying patient. Although the landmark volume focused almost exclusively on adult patients, its primary purpose, to transform patients into teachers of doctors, nurses, clergy, and their family members, also applied to her examination of children's voices in a later volume.[44]

Through *On Death and Dying,* a lecture tour, and an interview in *Life* magazine, Kübler-Ross's argument to restore agency to the sufferer, to replace technology with humanity, and to involve the patient in the process of death gained popular resonance. She became a leader in thanatology, a new field that studied concerns related to death.[45] The cover story of *Newsweek* in May 1978 described the "death-awareness movement" in America that reexamined taboos against death and revised the dehumanizing rituals identified in *The American Way of Death* a decade earlier.[46] On a national level, a panel of officials from the National Cancer Institute, the National Institute on Aging, and other federal agencies began to evaluate how research institutions dealt with dying. Famed cancer patients including Senator Hubert H. Humphrey discussed their illnesses publicly, making national appearances or recording their experiences in illness narratives. An article in *Time* the following month noted, "Once it becomes apparent that an illness is terminal, conventional medicine often seems unequipped, untrained and even unwilling to deal with death."[47]

The movement to modify the medicalization of death in America influenced the care of pediatric cancer patients. In a 1976 article published in the *Journal of Pediatrics,* the mother of a child with cancer recalled a movie that depicted a physician struggling to face his dying cancer patient. When he composed himself and walked into the room, the girl said to him, "I am not cancer, I am not leukemia, I am a person."[48] This dramatic episode suggested a need for physicians to reorient themselves from the disease to the patient and to better cope with the possibility of child death.

The desire to provide comfort, treat pain, and reduce the use of aggressive treatment during terminal illness contributed to the launch of the modern hospice movement.[49] In 1966, Florence Wald, a nursing educator at Yale University, had invited Kübler-Ross and Cecily Saunders to lecture about changes in the care of the terminally ill. The following year, Saunders founded Saint Christopher's Hospice in London. These key events helped initiate the first three American hospice projects, Connecticut Hospice in Branford, Connecticut, St. Luke's Hospice in New York City, and Hospice of Marin in California, and, by 1978, a National Hospice Organization had been formed to coordinate hospice facilities and speak on their behalf. The burgeoning American hospice movement reaffirmed Kübler-Ross's principles by relocating the place of death from the hospital or nursing home facilities to a hospice or home and reforming the manner of death to one that focused on aiding, not isolating, the patient. Nevertheless, the majority of hospice programs did not include pediatric patients in the 1970s because it was common for adults who were admitted to hospice to have cancer, to have a six-month prognosis, and to be looking toward palliation rather than cure. Children did not fit easily into this admission profile and the standard guidelines for care.

In the mid-1970s, a three-year, National Cancer Institute–sponsored study led by Ida Martinson, a registered nurse and director of research at the University of Minnesota School of Nursing, considered the viability of home care for terminally ill children, in particular those with cancer.[50] Martinson was aware of hospice programs that served adults and implored her colleagues, "For a dying child and his family, I believe an alternative is imperative—and now." She explained, "The hospital with all its technology and highly skilled personnel not only may not be essential in caring for the dying child, but it may well be an obstacle to the provision of appropriate care for such a patient."[51] By refocusing the emphasis from cure to comfort during the terminal stages of illness, she sought to minimize the suffering of young patients and their family members. At home, the child's treatment regimen could be simplified to include only the procedures needed to manage pain or other side effects. She had observed that most tests were futile by the end of life; they only traumatized the child further and added to the high costs of cancer treatment. The two primary goals of the project were to restore control over decision making to parents and their ill children and to limit the uncertainty of the final stages of cancer—areas that she believed medical professionals often handled improperly.[52]

In many ways, the University of Minnesota was an ideal location for Martinson's experimental "Home Care for the Dying Child" project. In Minnesota and the surrounding area, most pediatric cancer treatment involved participation in cooperative clinical trials and was administered at centrally located, major cancer treatment centers. Families residing in Minnesota, Wisconsin, and North Dakota repeatedly traveled as far as 400 miles to bring their children to centers in the Minneapolis–St. Paul metropolitan area. Martinson viewed hospitalization as a particularly traumatic experience for children and families because it forced parents to endure long commutes between hospital and home and often divided families between two locations. If parents felt that they could manage the death of their child at home, and the child expressed this wish, Martinson argued that death should take place at home—a place she associated with security, familiarity, and, she predicted, lower cost care.

Martinson refined her experimental model through firsthand experience with Eric Kulenkamp, a ten-year-old boy with acute leukemia.[53] Eric was in the final stages of his thirty-month illness and had expressed that he no longer wanted to go to the hospital for treatment. Previously, Eric had embraced the role of "professional patient," constantly asking about his condition and learning the routines of his treatment. Eric's mother, Doris, had been his partner in therapy. According to her own estimates, she had spent at least part of one out of every four days (more than 200 days total) with Eric in the hospital or at the outpatient clinic. Over the course of his illness, she had witnessed Eric undergo nearly 500 laboratory tests such as complete blood counts and regular platelet counts. Like Martinson, she had questioned whether this intense regimen was necessary when his health began failing. She suspected that her son was now just a part of a larger experiment. She later reflected: "Being subjects of research is not terribly comforting . . . I, and others, were a bit paranoid at times, wondering if what was done was actually for the sake of our loved one or for the sake of research. In reality, it is a combination of both."[54] After Eric voiced his last wishes, his parents agreed to participate in the home-care experiment with the hope that they could delay or, ideally, avoid further hospitalizations.

Under his parents' constant watch, Eric was able to remain at home. The public health nurse assigned to the case trained his parents to administer injections, watch for infections or bleeding in his mouth, and cope with common complications. He died at home in his own bed in the middle of the night just seventeen days, eight home visits, and five phone consultations after the initial meeting between Martinson and Eric's family. Satisfied that they had been able to provide adequate care for their son, Eric's mother, Doris, converted her diary entries from the time into an extended narrative. Martinson and the public health nurse added their perspectives. The final product, titled *Eric,* suggested that facilitating children's death at home was a worthwhile project that should be pursued on a larger scale.

In *Meri,* another extended illness narrative, David Wetzel, Meri's father, shared his feelings about his ten-year-old daughter's illness and death from acute leukemia. Wetzel wrote that he and his wife had not thought beyond her cycles of relapse, treatment, and remission until relapse began to dominate the other phases. Wetzel wrote that at the time her home care began, "she had lived eight years with leukemia and we felt that she was suffering as much from her treatment as from the disease."[55] Meri experienced extreme bone pain from the proliferation of white blood cells in her bone marrow and she had rapidly moved from over-the-counter painkillers to Demerol, a potent prescription drug. After reading about the Kulenkamp's experience, they decided to care for Meri at home. They integrated her care into the pattern of their home life and successfully established a less confrontational relationship with her physicians, writing, "For the first time we felt like partners in her medical care."[56] Like many families, the labor to prevent and regulate Meri's worsening pain proved to be their biggest challenge as they tried to gauge the strength and dosage needed for relief and to persuasively communicate this to Martinson and the consulting physician. Martinson's presence and advice during Meri's illness, at her death and memorial service, and afterward helped the family cope and grieve. Martinson and other nurses in the project referred to themselves as "advocates for the dying" as they recognized that children's age and specific needs separated their experience from societal expectations of death.[57] Parent participants showed their support for "Home Care for the Dying Child" by recording their reactions with health professionals, dedicating memorial gifts to the research fund at the University of Minnesota School of Nursing, and, like Eric's mother and Meri's father, sharing their experiences.[58]

The project quickly gained institutional support from faculty members at the University of Minnesota. Physicians referred children aged five weeks to seventeen years old to the research team; the majority had forms of childhood cancer. By November 1977, Martinson's research team had worked with twenty-nine families, of which twenty children had died—seventeen at home and three in the hospital. Their involvement with the families averaged 32.5 days, with a range of 2 to 104 days and included professional services such as home visits, telephone calls, and accompanying families to clinic appointments for an estimated cost per child of $1,000. In contrast, Martinson's review of comparable hospital charts showed that the average hospitalization for a child with cancer was 28.5 days with a cost of $4,480.[59]

Martinson's "Home Care for the Dying Child" project proved to be a partial success. Results showed that end-of-life care was less expensive and less invasive at home. Parents reported that providing care at home relieved their guilt that they could not or had not done enough for their dying child. Under this system, parents not only managed their child's medical care, but also were able to fulfill their child's unique requests for favorite foods, familiar toys, visits from playmates or favorite family members, or other special needs. Families were moved to record their positive experiences as participants in the project and testified that it was possible to facilitate a "good" death at home.

There were, however, also several major challenges. Expert nursing care was at the core of the project, yet when Martinson administered a statewide survey to assess the attitudes of Minnesota's registered nurses toward death and dying, she exposed a gulf between their attitude, education, and experience and the needs of her study. Nurses affirmed that alternative systems of nursing care for dying patients needed to be developed, but only 50 percent of respondents named "home" as the appropriate location for death. She also found that nurses had great difficulty coping with ill children; 61 percent reported feeling very uncomfortable when dealing with a dying child.[60]

At the time of Martinson's study, significant changes were occurring concerning the professionalization and practice of nursing. In 1974, a small cohort of nurses convened at a meeting of the Association for the Care of Children in Hospitals and decided to form a professional organization for subspecialists working in pediatric oncology. A year later, the Oncology Nursing Society was established and, in 1976, the Association of Pediatric Oncology was incorporated with a mission to support nurses involved in a rapidly changing field.[61] As the role of nurses evolved to one that required providing direct patient care, administering clinical trials and implementing research protocols, and working closely with young patients and families, this specialized group required a central association to link individuals who interested in the challenges posed by a rare set of diseases and, at times, tragic deaths. The need for a strong regional and national support system was evident in the challenges posed by Martinson's study.

At the same time, Martinson also encountered hostility from cancer specialists. She had received referrals from a number of physicians at the university and the surrounding region, but there was still profound ambivalence to hospice care. Donald Pinkel, a pediatric oncologist at St. Jude, wrote, "For many families the death of a child is better conducted at home or in a community hospital."[62] One of his colleagues countered, "Not everyone needs a hospice," labeling such efforts "anti-therapy and anti-therapist."[63] At a time in late twentieth-century medicine when cancer medicine was focused on exhausting all options before abandoning hope for a cure, choosing to stop could be met with criticism from physicians. Significant work would have to be done if collaborations like those built at the University of Minnesota were to be implemented widely.

In 1983, Children's Hospice International was founded, and by 1985, 183 hospices had opened their doors to children. It soon became apparent, however, that there was a serious mismatch between the vision of Martinson and others and the realities of pediatric cancer treatment in the United States. One frustrated hospice organizer spoke for the entire community when he explained, "Everyone looks so hard for remission that the child [with cancer] might be dying while receiving aggressive therapy."[64] Few children in hospice care had cancer because it was difficult to accurately predict the life expectancy of children with cancer, and many parents were unwilling to halt treatment even if given only a slim chance of saving their child's life. The tendency, hospice workers remarked, was to treat to the end. A redesigned, more flexible choice was needed to successfully accommodate all children with cancer, their parents, physicians, and insurers.[65]

Child Advocacy and Survivorship

In the 1970s, issues of health, education, justice, poverty, and abuse and their relationship to the treatment of children became the basis for critical discussion and action in America. Stephen Hess, the chairman of the White House Conference on Children, stated, "The child—as far as our institutions and laws are concerned—is too often a forgotten American."[66] In *The Children's Rights Movement,* Beatrice and Ronald Gross urged each person to become a "child advocate" who would speak out for this voiceless population.[67] In the 1970s, parents became vocal advocates through a small, yet significant cluster of illness narratives and the creation of a new national organization devoted solely to childhood cancers and the families affected by this set of diseases.

This phenomenon explains in part why parents increasingly wrote cancer narratives in the 1970s. Parents of childhood cancer sufferers—like other parents writing about hemophilia and cystic fibrosis—informed readers about the new course of these diseases.[68] By tracing their family's experiences during each stage of medical management from diagnosis, through their treatment decisions, to terminal care or possible long-term survival, the narratives acted as a comprehensive guide for other families facing cancer.[69] Ilana Löwy has suggested that cancer narratives commonly reflect the (sometimes reluctant) adherence of many cancer patients and their families to the oncologists' concepts and practices.[70] At the same time, accounts emphasized parents' involvement in managing their child's disease on their own terms. In some cases, they wrested control from health professionals and thought that their personal involvement in their child's care improved their child's hospitalization or healing process.[71] Barron Lerner has noted that a large collection of book-length illness narratives and breast cancer accounts in newspapers and popular magazines and on television served as powerful outlets for women to criticize and revolt against standard medical and surgical practices in the 1970s.[72] Childhood cancer narratives contained fewer overt criticisms of their child's care, but parents subtly interwove the personal and political in their published illness narratives as a way to participate in broader debates and public scrutiny toward medical care, the rights of children and other populations, and issues of death and dying that intersected in the experiences of all cancer patients and their families in the 1970s.[73]

Published in 1974, *Eric,* the story of a seventeen-year-old boy's experience with acute leukemia typifies childhood cancer narratives published in the 1970s and 1980s.[74] Many accounts highlighted the broad public awareness of cancer and the dread still associated with the disease. In the book's opening pages, Doris Lund, Eric's mother, described her poignant reaction to her son's diagnostic tests and diagnosis:

> I blanked out the words "bone marrow" instantly as if I'd never heard them before in all my life, as if I'd never read a single book or magazine article or watched a single TV drama which spelled out in the plainest possible terms exactly what a doctor was looking for when he ordered a bone marrow. After all, there was that perfect physical exam only twelve days before. . . .[75]

When the doctor called the next afternoon to schedule an appointment to share the diagnosis, Lund immediately replied, "You don't have to. I already know. Eric has leukemia."[76]

Eric, like most accounts, also detailed each step of the teenager's treatment. Massive doses of chemotherapeutic drugs were able to induce six consecutive remissions. Eric, though, was unusually sensitive to the drugs, and severe side effects such as mouth ulcers and nosebleeds prevented physicians from administering the full doses. As Lund observed, his physicians walked a "tightrope," a careful balance between inducing remissions and preserving healthy tissues and their vital functions. After his first remission was obtained, Eric traveled to Memorial's outpatient clinic from his home in southwestern Connecticut for weekly blood tests and regular bone marrow exams. Through maintenance therapy, a course of drug therapy administered at home during remission to continue the healthy status, Eric remained in his first remission for more than a year.

Narratives also illustrated how patients procured experimental drugs when standard therapies had been exhausted. After Lund read about asparaginase, an experimental drug available in limited quantities, she approached Joseph Burchenal, director of clinical investigation at Sloan-Kettering, personally. According to Lund, Burchenal, stated bluntly, "There are six patients who ought to be getting asparaginase right now. I think that we have enough in that bottle for two."[77] After suffering relapses, Eric was given injections of vincristine and prednisone three times per day. To his mother's amazement, he also received asparaginase once each day as an outpatient at Memorial. Because of the experimental nature of the drug, he had to remain at the hospital all day so that staff members could identify, monitor, and record its side effects.

Physicians decided to hospitalize Eric after finding that the extreme nausea stimulated by the asparaginase made it impossible for Eric to ride home in the car after chemotherapy. He was housed in the clinical research facility, a separate service at Memorial designated only for research subjects. Despite carrying no major medical insurance, the Lund family had paid for Eric's doctor and hospital costs. Upon his admission to the clinical research facility, a federal grant given to hospitals investigating experimental drugs now paid for his treatment. Lund wrote that no one would be equipped to afford the new drugs or the intensive nursing care and special diets that were required during testing.

Parents emphasized the complications that accompanied the cycles of remission and relapse and their child's slow decline as periods of health became shorter and more difficult to maintain. During acute episodes, family and friends needed to donate adequate units of blood to address the demands of Eric's illness. A centrifuge separated plasma and platelets from whole blood through plasmapheresis, so the red blood cells were returned and the donor could give again in only two to three days. Eric also developed a tumor behind his left eye that required pinpoint radiotherapy. Before his death, Eric suffered a severe crisis that resulted a four-day coma from infection and near organ failure.

Like most parents, Doris Lund had maintained hope that the next advances in chemotherapy would not only continue to prolong her son's life but provide a cure. As she and her son watched astronauts land on the moon, Lund had been amazed at the drama of the event:

> I thought, it was a miracle watching a miracle. It had taken billions of dollars to put those astronauts on the moon. It had taken millions in medical research to put Eric Lund, very much alive and well, in a Colorado tavern nearly two years after he was stricken with leukemia. Not too many years before, his life would have been over five or six weeks after diagnosis. Science had given him a gift of years. Because of that gift, Eric got to watch the greatest event of his time.[78]

Although she repeatedly professed her faith in science and denied that Eric would die, Lund and her family refused any life-saving interventions that may have prolonged his suffering. Only at the terminal stage of illness did parents accept the reality of their child's death. Using illness narratives, parents created a permanent record that preserved their child's ability to cope courageously with illness, gave lasting life to one cut painfully short, and served as a creative outlet to express grief.[79]

Childhood cancer narratives explored how illness altered their parent-child relationship and provided insight into the important influence of age on the sufferer's and the family's experience of illness.[80] Eric's age helped define the course of his disease, since young people over the age of twelve had poorer acute leukemia prognoses. His age also shaped his experience. As a seventeen-year-old, Eric wanted to contribute equally to the decisions regarding his care. He knew that he was seriously ill and factual information served to satisfy, not agitate him. Eric asked his mother for permission to manage his illness himself though private conversations with his doctor.[81] Eric became preoccupied with maintaining a positive attitude, fighting the disease, and staying out of the hospital as much as possible. During remissions, he returned to school for brief periods and was elected as captain of his collegiate soccer team. He began dating Mary Lou, his private duty nurse at Memorial, and gave her a copy of *On Death and Dying* as his death neared. During the last stage of his illness, he began to discuss how he wanted to die and signed forms in order to donate his eyes. Lund thoroughly documented the events that unfolded throughout her son's illness, but she also captured her own struggles to protect him while according him the independence of a young man.

A final set of narratives was written to alert Americans of the mounting evidence about hazardous environmental toxins and their detrimental effects on child health. Heightened awareness of the direct effects of environmental pollutants such as DDT and dioxin on the health of plants, animals, and humans followed the publication of *Silent Spring* in 1962 and the establishment of the Environmental Protection Agency.[82] The first major epidemiological investigation of a childhood leukemia cluster occurred in Niles, Illinois, in 1963.[83] A cluster of thirteen leukemia cases during a four-year period in the town initially raised concerns. The children of Niles had little in common—they were of different ethnicities, their parents did not have similar occupations, and radiation levels in the town were not abnormally high. The only common factor was that all of the affected children resided in Niles and attended or had attended the same Roman Catholic primary school. Trained U.S. Public Health Service investigators from its Communicable Diseases Center in Atlanta conducted a house-to-house study in the town in order to directly link the cases to a common genetic, chemical, or infectious factor, but no conclusive link was found.[84] In the decades that followed, the Centers for Disease Control investigated dozens of other cases including sites in the Love Canal area of Buffalo, New York; residential neighborhoods in Woburn, Massachusetts; and a school in Rutherford, New Jersey.[85] In these three areas, parents suspected that improper disposal of toxic chemical waste were responsible for their children's cancers. Two mothers, Anne Anderson and Lois Marie Gibbs, led grassroots campaigns to accurately count and plot the number of ill in their communities and to persuade public health officials that industrial dumping threatened the health of all of the town's citizens—especially the youngest residents.[86]

The Anderson family of Anne, her husband Charles, and children Christine, Charles Jr., and Jimmy became the center of a controversy over the health risks of toxic chemicals dumped in Woburn, Massachusetts, an industrial suburb located eleven miles north of Boston. At age three, Jimmy had been diagnosed with acute leukemia. Under the supervision of John Truman, head of pediatric oncology at Massachusetts General Hospital, he was treated with an intense treatment regimen pioneered at St. Jude Hospital that included radiation of the skull and combination chemotherapy. Jimmy and his mother traveled to Boston on nearly a daily basis for therapy and management of his short-term side effects. He also suffered permanent impairments such as a speech impediment, a learning disability, and a deficiency in fine motor control. At age seven, Jimmy relapsed. In *Cluster Mystery,* author Paula DiPerna wrote, "In this, the small boy's ordeal that was the life of Jimmy Anderson gradually became the chronicle of a town and a medical touchstone for a nation."[87]

Anderson kept an informal list of the cancer victims and their locations and recorded the cases on a pushpin map to vividly illustrate their number, location, and proximity to one another. Struck by her findings, she gathered a complete victim list by holding a city meeting. She found that between 1968 and 1979, twelve children from East Woburn were diagnosed with acute lymphocytic leukemia; six children lived close to one another in a geographical "cluster." Through FACE (For A Cleaner Environment), Anderson and other parents organized community meetings, distributed a newsletter, and tried to build critical connections between extensive toxic waste dumping, well contamination, and childhood leukemia.[88] On December 12, 1979, the headline "Child Leukemia Answer Sought" was the first public announcement of the cluster.[89] Throughout the case, a complicated relationship arose between expert scientific knowledge versus the public interest and the concerns of the individual.[90]

On May 22, 1979, wells G and H, two sources of water for Jimmy's home, were ordered closed immediately by the Department of Environmental Quality Engineering of the Environmental Protection Agency serving the state of Massachusetts because the water supply was contaminated with chemicals found to cause cancer in lab animals. As local newspaper articles unearthed details about illegal toxic waste dumping and hearings were held about the possible health effects of hazardous waste disposal, Anderson permitted reporters to interview Jimmy about the Woburn case. Patrick Toomey, another Woburn boy dying of cancer, spoke with Senator Kennedy about his experiences and the local situation.[91] On January 18, 1981, Jimmy died from a massive pulmonary hemorrhage caused by aplastic anemia, a side effect of chemotherapy. Jimmy and Patrick's deaths helped convince parents, other residents of Woburn, and those reading about the case in the *New York Times* and the *Washington Post* that the children's suffering was linked to improper toxic waste disposal by the Grace Corporation and Beatrice Foods. Based on a computer model of water distribution from the contaminated wells and the statistical disease evidence, eleven Woburn families filed a civil suit against the corporation. In 1984, study results found a positive association between toxic waste and birth defects, stillbirths, and childhood leukemia. A verdict found that Grace had negligently dumped chemicals on its property, and the out-of-court settlement that followed captured national attention because it demonstrated corporate responsibility for proper toxic waste removal and the attendant physical and psychological health risks from improper disposal. Young victims such as Jimmy Anderson and Patrick Toomey, their families, and other community activists identified and acted against corporations and local, state, and federal officials for the remediation of toxic waste problems and the recognition that their actions had caused human suffering.

Long-Term Survivorship

In 1970, a new national childhood cancer foundation joined a long list of organizations such as the Jimmy Fund, the Leukemia Society of America, the Bright Star Foundation begun by the Bush family, and the Association for Brain Tumor Research that was established by two Chicago mothers concerned about research for inoperable tumors.[92] Although some were started by parents, others like St. Jude Children's Research Hospital and the American Lebanese Syrian Associated Charities, its fundraising arm, were launched by celebrities. Danny Thomas, a performer known for his role on the television program *Make Room for Daddy*, used his links to the entertainment industry to stage large-scale fundraising events to benefit an institution focused narrowly on the research and treatment of pediatric cancers.[93] In the past, like today, the dozens of nonprofit and for-profit groups for pediatric cancers were started on the local, national, and international level to help fill perceived voids in cancer services. Although few are linked directly, many shared a common origin—the premature illness or death of a loved young child from cancer.

In 1968, the daughter of Grace Ann Monaco was diagnosed with acute lymphoblastic leukemia. During her daughter's illness, Monaco met parents of other children in treatment at Children's National Medical Center in Washington, D.C., and witnessed how childhood cancer and the daily demands of treatment affected the entire family. Monaco founded Candlelighters in 1970 to create a network of peer support and information for patients and their families, a goal that resonates with the wishes expressed by parents who wrote the Gunthers in the 1940s and early 1950s. In the decades that followed, parents and patients had contact with others in outpatient clinic waiting areas or in formal group meetings organized at individual research centers, but Candlelighters expanded communication beyond the institutional setting to local, regional, and national forums. Like parents of children with disabilities, members of Candlelighters sought to affect change in medical and social systems by gathering the voices of individual parents into a formal, specialized organization.[94]

By visiting children's hospitals across the country, Candlelighters' early members gained first-hand knowledge about the cancer-related concerns of children and met professionals willing to volunteer their time to help children. In 1970, the Childhood Cancer Ombudsman Program began organizing panels of volunteer doctors and lawyers willing to give free opinions on issues such as treatment choices, informed consent, employment discrimination against parents, education discrimination, and barriers to insurance. Candlelighters also promoted the need for more information about the proper nutrition for children with cancer. Children's size fluctuated at different points during treatment, but little was known about how to prevent these problems. Based, in part, on these efforts, the National Cancer Institute began a diet, cancer, and nutrition program that was first funded in 1974. The group also used their presence in Washington, D.C., to lobby legislators for access to pediatric clinical trials.

In 1978, Candlelighters convened "Maintaining a Normal Life," its first national gathering for sufferers, survivors, and parents of children with cancer. The number of childhood cancer groups associated with Candlelighters had increased from 3 to 100 groups in 47 states, Canada, and Europe, and approximately 400 adults, adolescents, and children from across North America participated in the conference. The National Cancer Institute's office of communications also recorded and published the conference proceedings so that they would be widely available.[95]

The conference program highlighted the growing challenges presented by cancer survivorship. Few of the invited speakers or sessions dealt directly with innovations in treatment regimens; rather, most were dedicated to issues related to the prolonged medical management of the disease and the long-term survival of children that characterized childhood cancers by the mid- to late 1970s. Jordan Wilbur described Stanford's ten-year history of completely integrating families into the daily activities of the oncology unit so that they could master skills such as dispensing oral or intravenous medication before the child returned home.[96] Other talks described the development of outpatient clinic services and the facilitation of children's reentry into "normal life" at home and school during treatment. The provision of adequate supportive care at home was a crucial subject for parents caring for a child on a multiyear chemotherapy protocol.

Questions and comments from parents and teenage sufferers demonstrated that combining medical care and "normal" life posed many problems for families. Discussion focused on relations within the patients' family, practical problems including discipline, nutrition, cost, and home care, and the child's or teen's perception of illness and death. At each session, preselected panels composed of parents responded to the speakers' presentations and began a dialogue that often reflected tensions between expert advice and personal experience. A panel of ten teenagers with a variety of common childhood cancers also shared their frank responses. They expressed frustration yet resignation toward multiple relapses, the acute nausea that accompanied chemotherapy, their frequent absences from school and social life, and the stigmatization that followed amputation or hair loss. At the same time, however, each youth told of their overwhelming determination for continued life and revealed that they were not content to be viewed as passive recipients of care.[97] The teenagers willingly shared their opinions among this small circle of peers, but they also self-consciously sought to extend their influence beyond the panel to interested parents, physicians, and other readers of the conference proceedings.

As all childhood cancer survival rates slowly improved in the 1970s, "late effects" of therapy—delayed physiological and psychological consequences of their aggressive, repeated surgery, radiation, and drug treatment—first became apparent. Parents attending the conference session on the outlook for children who had completed treatment had many unanswered questions. After enduring a prolonged illness alongside their children, parents worried that their child's normal growth and development would be permanently retarded by the side effects of their life-saving treatment. Nevertheless, the presenter could not make definitive statements based on the limited data that was available. Investigation into late effects and the future for a growing number of childhood cancer survivors escalated into the twenty-first century.

During the 1970s, young acute leukemia sufferers like Amy Louise Timmons spent years cycling between periods of health and illness. Therapeutic breakthroughs in the form of effective chemotherapeutic agents and complex chemotherapy protocols had transformed the course of this disease from a rapid killer to a treatable illness with long-term survivors. By this decade, comprehensive cancer centers became key sites for advanced childhood cancer treatment and many children spent years receiving regular outpatient care there after returning to home and school. Although care centered in the home enabled children to live in a familiar environment and reduced burdens of long-term hospitalization, it introduced new concerns related to maintaining daily family life while dealing with the side effects of toxic chemotherapeutic drugs and the unexpected exacerbations of leukemia. Illness narratives revealed young patients' courageous battles with illness while at the same time illuminating parents' struggle to care for their child while maintaining the family's home life. Parents gained additional responsibility as the primary caretakers of children with cancer: parents carefully monitored their children's health, questioned the use of experimental therapies in their ill and dying children, chose to supplement conventional care with alternative therapies, and dictated the terms of their children's deaths. As parent advocates, they participated in negotiations with professional health care providers and, sometimes, industrial polluters and the law. Despite numerous setbacks, guarded hope for a permanent cure united parents, physicians, and researchers.

At a 1977 conference on leukemias and lymphomas sponsored by the American Cancer Society and the National Cancer Institute, Joseph Simone, an oncologist at St. Jude, confidently stated, "It is no longer controversial to state that a significant proportion of children with acute lymphocytic leukemia can attain long-term, leukemia-free survival and, possibly, cure."[98] Simone reported that some of the young patients he treated at St. Jude had completed high school and college, married, and had children after regaining their health, but cures had not been achieved without a cost. As another speaker at the conference later noted, "The gratifying success of therapy permits the physician to divert some of his attention from cure to the consequences of the treatment he is giving."[99] The specter of late effects and their long-term consequences for children who had been cured of cancer through aggressive radiation and chemotherapy remained.

Conclusion

In October 1939, metropolitan newspapers reported that surgeons at Memorial Hospital had amputated the right leg of eight-year-old Dorothy Lewis in order to remove a malignant bone tumor completely.[1] Cancer specialists had advised her father that radical surgery held the only hope of completely removing the growth and predicted that Dorothy would survive only a year without the operation. Despite this dire prognosis, the girl's father, William Lewis, a laborer for the Queens Parks Department, repeatedly refused to permit the potentially life-saving operation, admitting, "I couldn't stand Dorothy's being a cripple."[2] Instead, he permitted his daughter to undergo weekly radiation therapy. Not long after her diagnosis, newspaper articles about Dorothy and the physician-parent dispute prompted readers from around the world to send letters to the Lewises' Brooklyn home to offer their support or to challenge the father's stance regarding his daughter's treatment.[3] When treatments failed to shrink the tumor or provide lasting symptomatic relief, Lewis allowed the amputation. After her surgery, physicians cautiously labeled Dorothy's condition "satisfactory" and reported that they believed the tumor had not spread to other parts of her body.[4] This announcement concluded the yearlong debate.

Thirty-four years later, Edward M. Kennedy, Jr., complained to his father that his lower leg was painful and swollen. The twelve-year-old, known as Teddy, was the son of Edward Kennedy, the chairman of the Senate Subcommittee on Health and Scientific Research. Their family physician initially dismissed Teddy's symptoms and recommended that he soak the leg in Epsom salts, but a second examination and biopsy by George Hyatt, a professor of surgery at Georgetown University Hospital, revealed a tumor. Hyatt recommended prompt amputation to prevent the rare, fast-growing cancer of the cartilage from spreading. Only four days after his diagnosis, Teddy underwent an hour-long operation to amputate his leg above the knee and to form a stump for attaching a prosthetic device. Brief articles in the *New York Times* updated readers on Teddy's treatment regimen, his release from the hospital and continued treatment every three weeks, his first ski trip after the surgery, and his correspondence with a penpal who was also an amputee. Teddy also received thousands of letters and telegrams wishing him a speedy recovery.[5]

After surgery, Teddy participated in a study at the Dana-Farber Cancer Center (previously the Jimmy Fund Clinic) that investigated the role of adjuvant chemotherapy in children with osteogenic sarcoma who had been treated with immediate amputation.[6] A pair of articles published in the *New England Journal of Medicine* reported that by adding chemotherapy to surgical intervention, clinicians could defer or prevent the onset of lung metastases. Research at the cancer center found that relapse was deferred with the administration of methotrexate and citrovorum factor.[7] To prevent dangerous side effects such as anemia, nausea, mouth ulcers, and impaired organ function caused by the methotrexate, the effective but toxic dose was followed by a massive dose of citrovorum factor; this combination allowed methotrexate to selectively target the tumor by sparing the normal cells that were "rescued" with the citrovorum factor antidote. As part of the study, Teddy stayed at the cancer center for six-hour continuous intravenous infusions of methotrexate in which he received up to 100 times the standard dose of the drug. Two hours later, the administration of citrovorum factor began and continued up to six days by oral dose. When the methotrexate level in blood had declined satisfactorily, he was allowed to return home for his body to recover. Citrovorum factor allowed researchers to continue this treatment for long periods of time without inducing life-threatening side effects in the young patients. Only two patients of the twenty-person study group developed lung metastases. Teddy was cured.

Immediate amputation may have prolonged Dorothy Lewis's life, but it did not guarantee a cure. The radiotherapy that temporarily stalled the tumor's growth and the delay before her surgery may have assured—not caused—a virtually inevitable death from lung metastases. It was not until the 1970s that chemical agents and the principle of citrovorum factor rescue—a protocol developed through acute leukemia and adjuvant chemotherapy research programs in the 1950s—had improved the chances of a long-term survival or cure for Teddy and other children with osteogenic sarcoma. However, it also extended the demands and duration of medical treatment for the young patient and his or her family.[8] The marked changes in the prognosis of bone cancer was reported in the medical literature and summarized in the popular press, but the newspaper articles that described Dorothy's and Teddy's cases vividly depicted and personalized new innovations. The stories of two children—one from a working-class family of five and one from a wealthy and politically powerful family—showed the contested nature of diagnosis and treatment that characterized many childhood cases from this entire period.

Both cases raise important questions about childhood cancer that this book has only begun to probe. Why did the specialists at Memorial allow Dorothy's father to delay her critical operation for more than nine months? Did this response indicate the degree of parents' influence at the child's bedside or physicians' uncertainty regarding treatment for bone cancers? Can it be traced to attitudes about physical disability, especially in children? It is difficult to know the answers based on the information given in the popular press, but Dorothy's physicians did appear guardedly hopeful about her recovery. Teddy Kennedy underwent surgery and was immediately enrolled in a clinical trial focused on a related but different cancer. His diagnosis may have been uncertain initially, but he may have also gained access to the experimental Dana-Farber study through his father's political connections and activities in shaping health legislation.

Why were these stories in the news? Teddy's story made the news, in part, because of the Kennedy's notoriety. But why did Dorothy Lewis's plight garner such sustained attention? The stories of individual cancer patients were often featured in the newspaper for a specific purpose such as soliciting donations for cancer hospitals or emphasizing the need for blood donors. In many cases, however, they seem to serve only as dramatic public interest stories. The series of articles and thousands of letters readers to Dorothy and Teddy attest to readers' fascination with the progress and outcome of children's cancer stories. Cancer, publicized as "the child killer," elicited a level of attention unmatched by many other childhood illnesses in the mid- to late twentieth century.

Dorothy Lewis and Teddy Kennedy joined "Jimmy," Johnny Gunther, and many others whose faces and chronicles were strategically used to raise awareness and funds for cancer research. *The Story of Teddy,* a movie about Teddy's illness, was aired on television. In a follow-up article after her surgery, Lewis was photographed while seated in a chair with her two favorite dolls. In the text, the reporter wrote that Dorothy had showed off her skill with crutches and hopped about the room, but the published newspaper photo masked her disability.[9] Like the youthful images and heartrending testimonials of American Cancer Society poster children, photos of Dorothy posed with her toys and Teddy on a ski trip emphasized the promise of improving survival rates and restoring children to full, healthy citizenship.

Abbreviated stories of sickness featured in popular articles, fundraising materials, and letters, as well as book-length narratives made the patient and the family—not the disease—the primary actors. By tracing changes in cancer awareness, research, and treatment through the pages of *Death Be Not Proud,* Angela Burns's letter, *The Blood of the Lamb, Eric,* and the cluster of illness narratives written in the 1970s and 1980s, we gain a more nuanced understanding of the toll cancer exacted from individual pediatric patients and their families. By valuing both medical accounts and stories of sickness, we approach a balanced knowledge of the changing course of many childhood cancers from the late 1930s through the 1970s.

A more complex image of childhood cancer research and treatment survival emerges by listening to children's and parents' own words. Only a portion of afflicted children received the first new chemotherapeutic agents popularly lauded as miracles because access often depended on proximity to a cancer center or the availability of adequate funds and transportation. Experimental therapies—especially those directed toward acute leukemia—promised new but, perhaps, false hope for children who had exhausted all other treatments. Chemotherapeutic agents threatened the health of the patient with intolerable, toxic side effects. In addition, they provided brief respites from the disease; a period of remission that prolonged children's lives from weeks to months to years but did not result in cures. This new pattern of disease challenged patients and families who returned to their normal daily activities but remained continually threatened with the unexpected exacerbations of an incurable disease.

Stories of individual sufferers also highlighted the communal aspect of cancer care. As concepts such as "total care" and "comprehensive cancer care" imply, the child was only one participant in the complex medical management that began to govern pediatric cancer treatment in the United States in the 1950s and 1960s. The child's local pediatrician or family physician, cancer specialists such as oncologists and hematologists, nurses, social workers, recreational therapists, and teachers were all a part of his or her multifaceted care—especially as cancer prognoses improved and children lived with the disease. Parents also played a prominent role in their children's treatment and became increasingly responsible for the daily demands of their children's illness after pediatric cancer care shifted to outpatient clinics and the home.

These duties tested marital relationships and placed burdens on families, yet parents also gained more time with and proximity to their children over the course of their illness and during their final days. Such concerns persisted as the number of pediatric patients treated in clinical trials continued to expand. According to the National Cancer Institute's figures, currently 55 to 65 percent of children in the United States who are diagnosed with cancer by or before the age of fourteen years enter a National Cancer Institute–sponsored clinical trial, as compared to 2 percent of adults.[10]

In their writing, parents not only illuminated the promise and limitations of cancer treatment, they also suggested the universal challenges posed by all catastrophic illnesses. In 1975, Robert and Suzanne Massie, parents of an eighteen-year-old boy with hemophilia, wrote, "The details of this struggle are personal, but the story itself is not unique. Every family with a handicapped or chronically ill child shares the same problems: lack of money, isolation from the community of the healthy, prejudice, misunderstanding in the schools, loneliness, boredom, depression."[11] They hoped that their account would impart strength to families faced with similar medical challenges and help readers who were unfamiliar with hemophilia better understand the disease. The Massies, like other parents and health activists in the 1970s, also used the book as a forum for voicing criticisms about the medical establishment: "Many parents felt trapped and silenced by the expertise and routine imposed by the clinic rules and accepted treatment."[12] Unlike the Massies and many adult cancer victims, most parents of young cancer sufferers expressed their disapproval in a less overt manner, however, most self-consciously depicted their child's cancer treatment as a series of negotiations between themselves, physicians, and allied health personnel. Denying or questioning experimental therapies, using alternative remedies, or providing terminal care at home all challenged medical authority over cancer care in children.

The end point of this history, the late 1970s, does not denote a definitive conclusion to the history of childhood cancer. New challenges faced professionals and parents responsible for children with cancer who were involved in long-term treatment plans or returned to home or school.[13] The establishment of specialized cancer camps, such as the Hole in the Wall Gang sponsored by actor and philanthropist Paul Newman, helped children find peer support in others who had cancer experiences and would not question their demanding treatment needs or hair loss.

Dramatic images and narratives about children with cancer and their families continue to attract attention at the local and national levels.[14] Alexandra

Scott, an eight-year-old girl who had been diagnosed with a neuroblastoma at infancy, died in August 2004. Years earlier, she and her family had started Alex's Lemonade Stand to raise money for pediatric cancer research and, specifically, for the two institutions who contributed to her care: Connecticut Children's Medical Center and Children's Hospital of Philadelphia. Widespread news coverage and appearances on popular television talk shows such as the *Oprah Winfrey Show* and the *Today Show* had rapidly disseminated her story to local and national audiences, and by the time of her death Alex and many other young volunteers had raised nearly $1 million through the construction of lemonade stands in all fifty states, Canada, and France.[15] The obituary by the Associated Press retold Alex's heroic story and included a photograph of the girl tending her drink stand, a simple wooden structure covered with handmade, rainbow-colored signs that asked for a fifty-cent donation and informed customers that the profits would benefit pediatric cancer research.[16] Framed by the stand, Alex sported a wide grin and a cocked hat that partially covered her bald head, a familiar marker for cancer. The jarring dissonance between a neighborhood lemonade stand—a common summertime activity for children—and the serious purpose of Alex's sales drew wide attention to her, the threat of childhood cancers, and the ongoing need for further research.

Alongside biomedical innovations like bone marrow transplantation and the identification of specific genetic and molecular markers of disease, pediatric oncologists have began to plot and study troubling patterns of "late adverse effects" among patients cured of cancer during childhood. New obstacles including secondary tumors and mental or physical impairment diminished the elation toward the promise and potential of cure. Since the late 1970s and early 1980s, pediatric patients, parents, and physicians have asked a series of new questions: Can survivors bear children? Do radiation treatments suppress bone growth or have permanent, deleterious effects on vital organs? Does chemotherapy's neurotoxicity cause motor dysfunction or learning problems? Are aggressive, lengthy chemotherapy regimens responsible for the occurrence of secondary cancers? Sites and systems of long-term follow-up care, new mechanisms for record keeping, and clinical research in this area has potential not only to answer these troubling questions but also to extend a child's "patient" status throughout his or her lifetime.

While child-focused efforts such as the Jimmy Fund and St. Jude's telethons continue to the present, the patient rights movement empowered adults with cancer (and their supporters) to share their personal experiences and critique their care.[17] An increasingly competitive, crowded field of health fundraising, marketing, and advocacy has refocused some attention away from children and rare childhood cancers and toward common adult cancers like those of the breast and colon that have responded well to adjuvant chemical and hormonal therapies.[18] While children's faces and tragic stories continue to garner particular attention and spur community action around individual sufferers, the term "poster child" has expanded widely to include spokespersons of all ages. The American public now not only expects that all children survive to adulthood; twenty-first-century health consumers expect scientific medicine to effectively prevent, manage, and, hopefully, eradicate chronic diseases like cancer throughout the human lifespan. It is our challenge to care for the young patients, survivors, and families who continue to struggle with this rare set of diseases in ways that honor their voices and best address their needs.

Introduction

1. Will Bradbury, "The All-Out Assault on Leukemia," *Life* 61/21 (18 Nov. 1966): 88–107.

2. On 26 Sept. 1942, a notice in the *Journal of the American Medical Association* announced that the Texas State Cancer Hospital project—a facility for cancer research and treatment—was underway with a $500,000 appropriation from the state legislature and a matching grant from the Anderson Foundation. By the time of Mike Parker's referral, M. D. Anderson had become a major cancer research center like Memorial Sloan Kettering in New York City. N. Don Macon, *Clark and The Anderson* (Houston: Texas Medical Center, 1976).

3. Bradbury, 94.

4. Bradbury, 106.

5. Alfred Rosenfeld, "A Superplan to Cut Years off the War," *Life* 61/21 (18 Nov. 1966): 108. The research program was designed to make hundreds of separate projects converge toward a single goal—the control and cure of leukemia.

6. Harold W. Dargeon, ed., *Cancer in Childhood and A Discussion of Certain Benign Tumors* (St. Louis: C. V. Mosby, 1940), 21–30. Although Dargeon persuasively argued for establishing a narrow, age-based definition of childhood cancer, I have expanded his range in order to capture the voices of children older than age fourteen who also suffered from common childhood cancers, although more rarely. Adolescents serve as valuable historical voices based on their ability to record their words, thoughts, and emotions in interviews, letters, and diary entries. I have attempted to consistently indicate children's ages or developmental stages with descriptive terms such as "infant," "child," or "adolescent" in order to highlight differences between age groups. It has also been my intent to highlight acute leukemia (a common cancer of the blood-forming tissues), but to also include the stories of children and families coping with the various types of common pediatric cancers. Finally, although available historical sources underscored the stories of children treated at specialized cancer centers, I have expanded this narrow story whenever possible in order to encompass the diverse experiences of families affected by childhood cancer.

7. Helen Hughes Evans, "Hospital Waifs: The Hospital Care of Children in Boston, 1860–1920" (Ph.D. diss., Harvard University, 1995).

8. Sydney A. Halpern, *American Pediatrics: The Social Dynamics of Professionalism, 1880–*

1980 (Berkeley: University of California Press, 1988), 57–59. See also Thomas E. Cone, Jr., *History of American Pediatrics* (Boston: Little, Brown, 1979).

9. Rima D. Apple, "Constructing Mothers: Scientific Motherhood in the Nineteenth and Twentieth Centuries." In Rima Apple and Janet Golden, eds., *Mothers and Motherhood: Readings in American History* (Columbus: Ohio State University Press, 1997), 90–110, and Julia Grant, *Raising Baby by the Book: The Education of American Mothers, 1800–1960* (New Haven, Conn.: Yale University Press, 1998).

10. Heather Prescott, *A Doctor of Their Own: The History of Adolescent Medicine* (Cambridge, Mass.: Harvard University Press, 1998).

11. Richard Meckel, *"Save the Babies": American Public Health Reform and the Prevention of Infant Mortality, 1850–1929* (Baltimore: Johns Hopkins University Press, 1990); Molly Ladd Taylor, *Raising a Baby the Government Way* (New Brunswick, N.J.: Rutgers University Press, 1986); Kriste Lindenmeyer, *A Right to Childhood: The U.S. Children's Bureau and Child Welfare, 1912–1946* (Urbana: University of Illinois Press, 1997).

12. James T. Patterson, *The Dread Disease: Cancer and Modern American Culture* (Cambridge, Mass.: Harvard University Press, 1987). See Walter Ross, *Crusade: The Official History of the American Cancer Society* (New York: Arbor House, 1987), for an institutional history of the organization.

13. Stephen P. Strickland, *Politics, Science, and Dread Disease: A Short History of United States Medical Research Policy* (Cambridge, Mass.: Harvard University Press, 1972), and Richard A. Rettig, *Cancer Crusade: The Story of the National Cancer Act of 1971* (Princeton, N.J.: Princeton University Press, 1977).

14. Scholars affiliated with the Centre for the History of Science, Technology, and Medicine at the University of Manchester have begun to focus on cancer efforts in Britain through a cancer-specific perspective. Under the direction of John Pickstone, Carsten Timmermann has explored lung cancer and other intractable tumors, Elizabeth Toon considered breast cancer and related education efforts, Helen Valier focused on the leukemias, and Emm Barnes studied childhood cancers. Together with David Cantor, they have formed an international community focused on the history and sociology of cancer. See *Special Issue: Cancer in the Twentieth Century, Bulletin of the History of Medicine* 81/1 (spring 2007), esp. John V. Pickstone, "Contested Cumulations: Configurations of Cancer Treatments through the Twentieth Century," 164–196, for an example of these efforts.

15. For example, some of the best work on the organization and activities of cooperative cancer groups has been written by sociologists Alberto Cambrosio and Peter Keating, "From Screening to Clinical Research: The Cure of Leukemia and the Early Development of the Cooperative Oncology Groups, 1955–1966," *Bulletin of the History of Medicine* 76 (2002): 299–334, and Ilana Löwy, *Between Bench and Bedside: Science, Healing, and Interleukin-2 in a Cancer Ward* (Cambridge, Mass.: Harvard University Press, 1996).

16. See Emm Barnes, "Caring and Curing: Paediatric Cancer Services since 1960," *European Journal of Cancer Care* 14/4 (Sept. 2005): 373–380.

17. Robert William Kirk, *Earning Their Stripes: The Mobilization of American Children in the Second World War* (New York: P. Lang, 1994). Other historians of childhood such as Steven Mintz, *Huck's Raft: A History of American Childhood* (Cambridge, Mass.: Harvard University Press, 2004); Paula S. Fass and Mary Ann Mason, *Childhood in America* (New York: New York

University Press, 2000); Harvey J. Graff, ed., *Growing Up in America: Historical Experiences* (Detroit: Wayne State University Press, 1987), and N. Ray Hiner and Joseph M. Hawes, *Growing Up in America: Children in Historical Perspective* (Urbana: University of Illinois Press, 1985), have also documented children's role in shaping American culture.

18. Russell Viner and Janet Golden, "Children's Experiences of Illness." In Roger Cooter and John Pickstone, eds., *Medicine in the Twentieth Century* (Amsterdam: Harwood, 2000), 575–587.

19. Physicians have written two notable books on the history of pediatric cancers. For the perspectives of scientific pioneers working in the field, see John Laszlo, *The Cure of Childhood Leukemia: Into the Age of Miracles* (New Brunswick, N.J.: Rutgers University Press, 1995), and Emil J. Freireich and Noreen A. Lemak, *Milestones in Leukemia Research and Therapy* (Baltimore: Johns Hopkins University Press, 1991). Childhood cancer sufferers and survivors—a small, yet exceptional fraction of cancer victims—have, however, remained virtually invisible in the historical record.

20. For insightful studies of these two other related groups, see Barron H. Lerner, *Breast Cancer Wars: Hope, Fear, and the Pursuit of a Cure in Twentieth-Century America* (Oxford: Oxford University Press, 2001), and *When Illness Goes Public: Celebrity Patients and How We Look at Medicine* (Baltimore: Johns Hopkins University Press, 2006). Also see *Bittersweet: Diabetes, Insulin, and the Transformation of Illness* (Chapel Hill: University of North Carolina Press, 2003), a recent history of juvenile diabetes, in which Chris Feudtner used a rich collection of patient records and letters from the Joslin Clinic in Boston to examine the changing experience of these young patients and their families as insulin therapy and diet regulation transformed diabetes from an acute to a chronic disease that required extended medical management.

21. Susan Resnick, a public health expert formerly affiliated with the National Hemophilia Foundation, divided the history of hemophilia into the following eras: the Dismal Era, before 1948, the Years of Hope, 1948–1965, the Golden Era, 1965–1982, and the AIDS Era, 1982–1988, in *Blood Saga: Hemophilia, AIDS, and the Survival of a Community* (Berkeley: University of California Press, 1999), 2. It is useful to characterize the history of acute leukemia by similar time periods and titles, although the last should be renamed to reflect the new realization that long-term survivors sometimes suffered from developmental delays and secondary tumors from their treatment.

22. John Gunther, *Death Be Not Proud: A Memoir* (New York: Harper and Brothers, 1949).

23. Angela Burns to John Gunther, 29 Jan. 1949, Box 45, Folder 2, John Gunther Collection, Special Collections, Joseph Regenstein Library, University of Chicago, Illinois.

24. Peter De Vries, *The Blood of the Lamb* (Boston: Little, Brown, 1961).

Chapter One • "Glioma Babies," Families, and Cancer in Children in the 1930s

1. Helaine Judith Colan, another child diagnosed with glioma, was called "America's most famous baby of the moment" and the "death or blindness" child in "History of the Making," *Los Angeles Times,* 15 May 1938, A5.

2. Patrice Pinell, "Cancer." In Roger Cooter and John Pickstone, eds., *Companion to Medicine in the Twentieth Century* (London: Routledge, 2003), 673.

3. The two French theses were C. J. Duzan, "Du Cancer chez les Enfants," *Thèse de Paris* (1876), and C. Picot, "Les Tumeurs malignes des Enfants," *Rev. Méd. De la Suisse romande* (1883).

See L. Haynes Fowler, "Malignant Epithelial Neoplasms, Carcinoma and Epithelioma, Occurring in Persons under Twenty-Six Years of Age," *Surgery, Gynecology, and Obstetrics* 43 (1926): 73, for a review of the Mayo Clinic cases.

4. Roger Williams, *The Natural History of Cancer* (New York: W. Wood, 1908.)

5. Physicians observed cancers in many of the same locations as common adult cancers, though the incidence was low. For example, these physicians reported isolated cases of ovarian cancer. Thomas H. Lanman, "Ovarian Tumors in Childhood with Report of Five Cases," *New England Journal of Medicine* 201 (1929): 555–562; Alexander A. Levi, "Malignant Tumor of the Ovary Occurring in a Thirteen-Year-Old Girl," *New England Journal of Medicine* 217 (1937): 595–597.

6. Ernest Scott, Marguerite G. Oliver, and Mary H. Oliver, "Sympathetic Tumors of the Adrenal Medulla: With Report of Four Cases," *American Journal of Cancer* 17 (1933): 396–433; John L. Redman, H. A. Agerty, O. F. Barthmaier, and H. Russell Fisher, "Adrenal Neuroblastoma: Report of a Case and Review of the Literature," *American Journal of Diseases of Children* 56 (1938): 1097–1112.

7. George Pack and Robert LeFevre. "The Age and Sex Distribution and Incidence of Neoplastic Diseases at the Memorial Hospital," *Journal of Cancer Research* 14 (1930): 167–294.

8. Ibid., 171.

9. Ibid., 216. Neoplasm was an early word for cancer that meant a benign or malignant new growth.

10. Harold Dargeon, "Pediatrics at Memorial Hospital for Cancer and Allied Diseases," Sloan Kettering Cancer Center papers, Rockefeller Archive Center, Pocantico Hills, New York, (hereafter referred to as SKCC papers, RAC).

11. Ibid.

12. Harold Dargeon, ed., *Cancer in Childhood and a Discussion of Certain Benign Tumors* (St. Louis: C. V. Mosby, 1940), 17.

13. Ibid., 14.

14. "Doomed Baby Is Barricaded in Home, Parents Bar Physicians as Jurists Act," *Herald Statesman,* 13 April 1933, 1.

15. Ibid., 2, "Father Leaves Baby's Fate to Court," *Herald Statesman,* 14 April 1933, 1; "Father Gives Fate of Child to Court," *New York Times,* 15 April 1933, 15.

16. See Emily K. Abel, *Hearts of Wisdom: American Women Caring for Kin, 1850–1940* (Cambridge, Mass.: Harvard University Press, 2000), and Charles E. Rosenberg, *The Care of Strangers: The Rise of America's Hospital System* (New York: Basic Books, 1987), to see the transition in the location of care.

17. Judith Walzer Leavitt, *Typhoid Mary: Captive to the Public's Health* (Boston: Beacon Press, 1996).

18. Berthold Lowenfeld, former superintendent of the California School for the Blind in Berkeley, California, described the changing social status of the blind in *The Changing Status of the Blind: From Separation To Integration* (Springfield, Ill.: Charles C Thomas, 1975). Legislation began to address the particular needs of the blind in the 1930s. In 1935, Title X of the Social Security Act included special provisions for assistance to the needy blind. The passage of other acts of legislation gave the blind preference for operating government-supported vending stands and provided funds for the Library of Congress to provide books for this population. In

1940, the National Federation of the Blind organized to represent the needs of the blind to legislators in order to establish the rights of the blind and to promote their equal status with those with sight; however, it was the recognition of retrolental fibroplasia in premature infants caused by oxygen therapy and the outbreak of rubella epidemics that stimulated a comprehensive set of educational services for blind children in the 1940s, 1950s, and 1960s.

19. S. Finestone, *The First Thirty Years: A History of the National Federation of the Blind* (Des Moines, Iowa: National Federation of the Blind, 1971), 12–13

20. LeRoy Ashby, *Endangered Children: Dependency, Neglect, and Abuse in American History* (New York: Twayne, 1997).

21. The first Society for the Prevention of Cruelty to Children was organized in New York in 1875 to press for the reform of the criminal law surrounding the social regulation of parent-child relations and to develop an agency that could investigate cases of suspected cruelty to children. See John Macnicol, "Welfare, Wages, and the Family: Child Endowment in Comparative Perspective, 1900–1950." In Roger Cooter, ed., *In the Name of the Child: Health and Welfare, 1880–1940* (London: Routledge, 1992), 147, and E. Marguerite Gane, "A Decade of Child Protection," *Annals of the American Academy of Political and Social Science* 212 (Nov. 1940): 153–158, for a discussion of the primary principle of the welfare program—the right of protection from neglect and abuse.

22. W. L. Benedict, "Retinoblastoma in Homologous Eyes of Identical Twins," *Archives of Ophthalmology* 2 (1929): 545. Articles discussed two possible causes of retinoblastoma: heredity or spontaneous growth. Parents who had lost children from the disease frequently asked whether they should have more children, and those whose children survived questioned whether their children should have offspring.

23. Reese began his training in the pathology of the eye in Boston and then completed a two-year residency at the New York Eye and Ear Infirmary before moving to Vienna to spend a year studying with the eminent ophthalmologist Ernst Fuchs. Upon returning to the United States, he organized a department of ocular pathology at the Institute of Ophthalmology at Presbyterian Hospital in New York. See Algernon B. Reese, "Extension of Glioma (Retinoblastoma) into the Optic Nerve," *Archives of Ophthalmology* 5/2 (Feb. 1931): 269–271. See also Alan C. Wood, "Algernon Beverly Reese: An Appreciation," *American Journal of Ophthalmology* 71/1 (Jan. 1971): 137–142, and, secondarily, Ira Snow Jones, "Algernon Beverly Reese," *American Journal of Ophthalmology* 71/1 (Jan. 1971): 143–150. While tumors were recognized as a rare yet serious threat to sight, much more attention was devoted to widespread medical and public health problems such as trachoma, congenital cataracts, the harmful results of vitamin deficiencies, and the importance of systematically preventing ophthalmia neonatorum by administering drops of silver nitrate at birth. See Philip A. Halper, "The Challenge of Ophthalmologic Problems: Do We Meet Them?" *The Sight-Saving Review* 3/1 (March 1933): 15.

24. It was later shown that the tumor did not move from one eye to the other through the optic nerve; each tumor originated separately. Also, autopsy results reported in the literature and conducted at the Institute of Ophthalmology, Memorial Center for Cancer and Allied Diseases demonstrated that the tumor was not fatal only by extension into the intracranial space. In more than 50 percent of the cases death was due to metastasis to the viscera and distal bones. Algernon B. Reese, *Tumors of the Eyes* (London: Cassell, 1951), 85.

25. In *Tumors of the Eyes,* Reese wrote that when both eyes were affected and the child had

already lost his or her sight, he allowed parents to choose whether the eyes should be removed. Nevertheless, he still thought that surgery was the best course. He stated, "Leaving the eyes in place until their removal is obligatory accomplishes no useful purpose, and only permits the certain wide dissemination of the growth, whereas life might be saved by immediate bilateral enucleation" (108).

26. "Father Leaves Fate of Baby up to Court," *Herald Statesman,* 14 April 1933, 13.

27. "Vasko Family with Sick Child Disappear," *Evening Star,* 15 April 1933, 1.

28. See Nancy Tomes, "Epidemic Entertainments: Disease and Popular Culture in Early Twentieth-Century America," *American Literary History* 14/4 (2002): 628, and Bert Hansen, "America's First Medical Breakthrough: How Popular Excitement about a French Rabies Cure in 1885 Raised New Expectations for Medical Progress," *American Historical Review* 103/2 (April 1998): 373–418, for insightful analysis of the relationship between newspapers, sufferers, and disease. For a more modern account of patients in the public eye, see Barron H. Lerner, *When Illness Goes Public: Celebrity Patients and How We Look at Medicine* (Baltimore: Johns Hopkins University Press, 2006).

29. "Doomed Baby Is Barricaded in Home," 1.

30. "Vaskos Hide Baby in Secret Refuge; Vanished Family Leaves No Trace, Thwarting Pursuit," *Herald Statesman,* 17 April 1933, 16.

31. "Doomed Baby Goes Blind As Family Flees," *Herald Statesman,* 19 April 1933, 3.

32. "Parents Flee with Ill Child," *Los Angeles Times,* 16 April 1933, 3.

33. The transcript of the decision was summarized in "Court Decrees Knife for Baby; Appellate Judges Uphold Order to Save Vasko Child," *Herald Statesman,* 18 April 1933, 1, and printed in full in "Vaskos Disappear as Baby Goes Blind," *Herald Statesman,* 19 April 1933, 3, and "Operation on Baby Ordered on Appeal," *New York Times,* 19 April 1933, 36.

34. Ibid.

35. Editorial, "Wise Daniels Come to Judgment," *Herald Statesman,* 20 April 1933, 10.

36. "Mother Yields Baby to Knife Today," *Herald Statesman,* 25 April 1933, 3.

37. "Vasko Child's Eye Is Removed Here," *New York Times,* 26 April 1933, 17.

38. Ibid.

39. "Eye Operation Performed on Vasko Baby; Condition Is Called Satisfactory," *Evening Star,* 26 April 1933, 1, 5; "Vasko Baby Gains, Plays with Dolls," *Herald Statesman,* 27 April 1933, 1.

40. In an earlier case, *The People of the State of New York v. Pierson* (1903), the court of appeals held that a parent or guardian may be convicted who, when a child under his care is dangerously ill, willfully neglects to summon to its aid a physician, that is, a person who is duly admitted to practice medicine under the laws of the state. Section 288 of the New York Penal Code provided that a person who "willfully omits, without lawful excuse, to perform a duty by law imposed upon him, to furnish food, clothing, shelter or medical attendance to a minor . . . is guilty of a misdemeanor." A law review points to this case as the first case in which such a question was raised in the U.S. in a court of higher resort, though other commentaries point to cases from the late nineteenth century. See "Failure to Furnish Medical Attendance to a Minor," *Columbia Law Review* 3/8 (Dec. 1903): 574–576. The Vasko case (*In re Vasko,* 263 N.Y.S. 552 [App. Div. 1933]) has often been cited as the earliest example of the state legally compelling parents to allow a recommended surgical procedure. Other key decisions that followed included *In re*

Rotkowitz, 25 N.Y.S. 2d 624 (Dom. Rel. Ct. 1941), ordering operation to correct child's foot deformity; *State v. Perricone,* 181 A.2d 751 (N.J. 1962), upholding guardian's authority to consent to blood transfusion for infant over parents' religious objections; *Jehovah's Witnesses v. King County Hosp.,* 278 F. Supp. 488 (W.D. Wash. 1967), holding that the state may intervene in parents' religiously motivated decision to refuse a medically necessary blood transfusion for their child); *In re Sampson,* 317 N.Y.S.2d 641 (Fam Ct. 1970), ordering surgery against parents' wishes to correct facial deformity; *Custody of a Minor,* 379 N.E.2d 1053 (Mass. 1978), ordering chemotherapy despite parents' pessimism and concerns about child's discomfort; *Petra B. v. Eric B.,* 265 Cal. Rptr.342 (Ct. App. 1989), ordering medical treatment for child's serious burns despite parents' desire to treat with herbal remedies; *In re Doe,* 418 s.E.2d 3, 7 n.6 (Ga. 1992), commenting that parents do not have an "absolute right to make medical decisions for their children"; and *A.D.H. v. State Dep't of Human Res.,* 640So. 2d 969 (Ala. Civ. App. 1994), ordering AZT treatment for HIV despite mother's insistence that her child was not infected with HIV).

41. "Vasko Baby Returns Home in Arms of Joyful Mother; New Case Up in Brooklyn," *The Herald Statesman* (Yonkers, N.Y.), 6 May 1933, 2.

42. Raymond S. Duff and A. G. M. Campbell, "Moral and Ethical Dilemmas in the Special-Care Nursery," *New England Journal of Medicine* 289 (1973): 885, and "On Deciding the Care of Severely Handicapped or Dying Persons: With Particular Reference to Infants," *Pediatrics* 57 (1976): 487.

43. For two recent examples, see Hazel Glenn Beh and Milton Diamond, "An Emerging Ethical and Medical Dilemma: Should Physicians Perform Sex Assignment on Infants with Ambiguous Genitalia?" *Michigan Journal of Gender and Law* 7/1 (2000): 1–63, and Alyssa Connell Lareau, "Who Decides? Genital Normalizing Surgery on Intersexed Infants," *Georgetown Law Review* (2003): 129–151.

44. "Battle to Save Life of Baby," *Chicago Daily Tribune,* 7 May 1938, 1.

45. Ibid. See also "Parents Prefer Baby's Death to Lifetime of Blindness," *Los Angeles Times,* 7 May 1938, 1; "Baby's Death or Loss of Sight is 'Left to God' by Its Parents," *Washington Post,* 7 May 1938, XI; and "Parents Debate Dooming of Baby," *New York Times,* 7 May 1938, 32.

46. Martin S. Pernick, *The Black Stork: Eugenics and the Death of "Defective" Babies in American Medicine and Motion Pictures since 1915* (New York: Oxford University Press, 1996). Pernick noted the reappearance of the Bollinger case in the 1938 coverage.

47. "Blindness Is Lesser Evil than Letting Baby Die, Voters Say," *Washington Post,* 19 June 1938. Although 63 percent of those polled were in favor of the operation, opinions varied according to religious affiliation. Seventy-three percent of Roman Catholics said they would choose to operate, 63 percent of Protestants agreed, while only 58 percent of nonmembers made the same decision.

48. "Surgeons Hopeful of Saving Baby from Total Blindness," *Los Angeles Times,* 11 May 1938, 1.

49. "Doctors to Fix Baby's Fate," *Chicago Sunday Tribune,* 8 May 1938, 1.

50. For the Canadian context, see Charles Hayter, *Element of Hope: Radium and the Response to Cancer in Canada, 1900–1940* (Montreal: McGill-Queen's University Press, 2005). See also Patrice Pinell, *The Fight against Cancer: France, 1890–1940* (London: Routledge, 2002), and Caroline Murphy, "A History of Radiotherapy to 1950: Cancer and Radiotherapy in Britain 1850–1950 (Ph.D. diss., University of Manchester, 1986). For a broad, comparative review, see John V.

Pickstone, "Contested Cumulations: Configurations of Cancer Treatments through the Twentieth Century," *Bulletin of the History of Medicine* 81/1 (spring 2007): 164–196.

51. "Doctors Strive to Save Baby's Sight; Eye Removal Halts Tumor's Fatal Spread," *Chicago Daily Tribune,* 10 May 1938, 1.

52. "Surgeons Remove Left Eye of Baby," *New York Times,* 10 May 1938, 3.

53. "Mother Talks: I Pleaded for My Baby's Life," *Chicago Daily Tribune,* 12 May 1938, 8.

54. "Colan Baby's Eye May Keep Sight," *New York Times,* 13 Aug. 1938, 15.

55. "Parents Prohibit Operation on Baby, Court Decree Sought," *Herald Statesman,* 12 April 1933, 2.

56. "Issues Stirred by Vasko Case," *Los Angeles Times,* 30 April 1933, 24.

Chapter Two • "Cancer, The Child Killer"

1. A recording of the original broadcast can be found on the Jimmy Fund website at www.jimmyfund.org/abo/broad/jimmybroadcast.asp. See also Saul Wisnia, *Images of America: The Jimmy Fund of the Dana-Farber Cancer Institute* (Charleston, S.C.: Arcadia, 2002), 18–19, for a rough transcript.

2. Wisnia, *Images of America,* 19.

3. George E. Foley, ed., *The Children's Cancer Foundation: The House that "Jimmy" Built, The First Quarter-Century* (Boston: Dana-Farber Cancer Institute, n.d.), 43.

4. Statistics from Memorial Hospital showed that only 18 of 218 patients with childhood cancers at the institution had survived at least five years from the date of admission. Harold Dargeon, ed., *Cancer in Childhood and a Discussion of Certain Benign Tumors* (St. Louis: C. V. Mosby, 1940), 17.

5. Ibid., 25.

6. U.S. Bureau of Vital Statistics, 1947.

7. Harold Dargeon, "Pediatrics at Memorial Hospital for Cancer and Allied Diseases" (1967), SKCC papers, RAC. In the years after the Registry's inception, its records showed rising rates of childhood cancer mortality. Dargeon attributed the trend to greater accuracy in diagnosis, reporting, and record keeping.

8. *The Bridge League Bulletin* (Nov. 1939), 1939–1940 Scrapbook, SKCC papers, RAC; "First Child-Cancer Ward Functioning in New York," *Newsweek* (15 Jan. 1940), 1939–1940 Scrapbook, SKCC papers, RAC.

9. For example, see *[Bloomfield, New Jersey] Independent Press,* 1939–1940 Scrapbook, SKCC papers, RAC, and photo, no date, RG 400.1, Series 4, Box 1 Public Affairs—Photos Department, Folder 45, SKCC papers, RAC.

10. Halloween–Children's Ward, stamped as Feb. 1949, RG 400.1, Series 4, Box 1, Folder 43, MSKCC Archives, RAC; Christmas–Children's Ward, 1949, RG 400.1, Series 4, Box 1, Folder 36, MSKCC archives, RAC; Christmas party, 1945, RG 400.1, Series 4, Box 1, Folder 31, RAC; Christmas–Children's Ward, no date, RG 400.1, Series 4, Box 1, Folder 31, RAC.

11. Easter Day party, 1949, RG 400.1, Series 4, Box 1, Folder 40, RAC.

12. Memorial Center for Cancer and Allied Diseases, *Quadrennial Report, 1947–51* (New York, 1951), 67.

13. Ibid.

14. Harold Dargeon, "Cancer in Children from Birth to Fourteen Years of Age," *Journal of the American Medical Association* (14 Feb. 1948): 459. The American Medical Association and the American Cancer Society jointly published this piece as part of a nine-part series on cancer.

15. "Children in Danger," *Newsweek* 29 (10 March 1947): 54,

16. "Tower of Hope," *Reader's Digest* (Sept. 1949): 13. This article was condensed from the 27 June 1949 issue of *Time.*

17. "Tower of Hope," *Time* (27 June 1949).

18. For a description of the growth of medical knowledge about leukemia, see Frederick W. Grant, "The Dread Leukemias and the Lymphomas: Their Nature and Prospects." In Maxwell Wintrobe, ed., *Blood, Pure and Eloquent: A Story of Discovery, of People, and of Ideas* (New York: McGraw-Hill, 1980), 511–546.

19. L. Emmett Holt, *The Diseases of Infancy and Childhood: For the Use of Students and Practitioners of Medicine* (New York: D. Appleton, 1897).

20. Ibid., 14.

21. Memorial Center for Cancer and Allied Diseases, *Quadrennial Report, 1947–51* (New York, 1951).

22. Despite Memorial's assurances, the desperation of cancer patients and the enthusiasm of scientists to apply new therapies to human cancers resulted in a transparency between treatment and research at Memorial and other similar cancer research centers. As in messages about adult cancers, dual messages about dread and hope frequently coexisted in discussions of children's cancers and justification for controversial research and treatment methods. Experimental cancer treatment and trials in children will be discussed further in subsequent chapters.

23. See Walter Ross, *Crusade: The Official History of the American Cancer Society* (New York: Arbor House, 1987), for an institutional history of the organization.

24. See James T. Patterson, *The Dread Disease: Cancer and Modern American Culture* (Cambridge, Mass.: Harvard University Press, 1986), for the best general account of the history of cancer in the United States. Patterson did not discuss children and cancer at length, but included chemotherapy and its successful use in acute leukemia in his chapter "The Research Explosion."

25. Bernard Behrend, "Do You Fear Cancer?" *Hygeia* 19/9 (Sept. 1941): 712–713, 752.

26. Surveys of the *Reader's Guide to Periodical Literature* and the *New York Times Index* reveal a steadily growing number of general articles about cancer in the 1940s and 1950s, but few writings on children's cancers were published. See Patterson, *The Dread Disease,* for figures from the 1940s and 1950s: "Between 1943 and 1945 the *Reader's Guide* contained 53 articles on cancer. By 1945–1947 the number had risen to 113, and by 1955–1957 to a peak of 273. Similar increases in the number of stories about cancer—from 54 in 1940 to 144 in 1955—occurred in the *New York Times.* Not even polio attracted so much special attention in the press at the time" (97, 143).

27. Frank Rector, "Cancer Kills Children Too!" *Women's Home Companion* (March 1947): 36–37, and Lawrence Galton, "Cancer, the Child Killer," *Colliers* (15 May 1948): 64–66.

28. Galton, 66.

29. Emily K. Abel, *Hearts of Wisdom: American Women Caring for Kin, 1850–1940* (Cambridge, Mass.: Harvard University Press, 2000). Also see Rima Apple, *Mothers and Medicine: A Social History of Infant Feeding* (Madison: University of Wisconsin Press, 1987), and Judith Walzer Leavitt, *Brought to Bed: Childbearing in America, 1750–1950* (New York: Oxford University Press, 1986).

30. In *American Pediatrics: The Social Dynamics of Professionalism, 1880–1980* (Berkeley: University of California Press, 1988), Sydney Halpern examined how pediatrics developed into an organized professional unit within American medicine between the 1880s and 1940s and then splintered into subspecialties after World War II. Other general accounts include Heather Munro Prescott, *A Doctor of Their Own: The History of Adolescent Medicine* (Cambridge, Mass.: Harvard University Press, 1998); Thomas E. Cone, Jr., *History of American Pediatrics* (Boston: Little, Brown, 1979); and Jeff Baker, "Women and the Invention of Well Child Care," *Pediatrics* 94 (1994): 527–531.

31. Nancy Tomes, *The Gospel of Germs: Men, Women, and the Microbe in American Life* (Cambridge, Mass.: Harvard University Press, 1998).

32. In *Generation of Vipers* (New York: Farrar and Rinehart, 1942), Philip Wylie coined the term "momism" to link mothers and psychological problems of youth, esp. boys. See Molly Ladd-Taylor and Lauri Umansky, *"Bad" Mothers: The Politics of Blame in Twentieth-Century America* (New York: New York University Press, 1998).

33. See Kathleen W. Jones, *Taming the Troublesome Child: American Families, Child Guidance, and the Limits of Psychiatric Authority* (Cambridge, Mass.: Harvard University Press, 1999).

34. Ladd-Taylor and Umansky, 4.

35. See "Navy Is Santa to Sick Child," *New York Times* (24 Dec. 1944): 21, "Victim of Leukemia Dies: Jersey Girl, 2, Won Sympathy of Many in Her Plight," *New York Times* (17 Nov. 1945): 32.

36. Stephanie Coontz, *The Way We Never Were: American Families and the Nostalgia Trap* (New York: Basic Books, 1992), 24.

37. "President Heeds Appeal: Refers to Army Case of Soldier Whose Baby Is Ill," *New York Times* (6 April 1944): 25; "Sergeant Truax on Way Home," *New York Times* (28 April 1944); "Soldier at Sickbed: Sergeant Home From Pacific Hopes for Child's Recovery," *New York Times* (30 April 1944): 28.

38. "Death Claims Baby of Sergeant Truax: Child Dies of Leukemia 6 Days after Father Returns on Leave," *New York Times* (5 May 1944): 21.

39. Galton, 64.

40. "Officer in from Germany Sees Son First Time—in Cancer Ward," *Herald Tribune* (19 Jan. 1946), Memorial Hospital 1946 Scrapbook, SKCC papers, RAC.

41. Galton, 65.

42. "Too Much to Bear," *Time* (15 Jan. 1951), 40. The Purcells faced a similar decision to the Vaskos and Colans. Their young daughter, Carolyn, had been diagnosed with retinoblastoma. Their home in Alpharetta, Georgia (population 647), was described as "small" and "dingy," and the article reported that Carolyn's father was an unemployed stonemason. The Purcells were offered surgery at Mayo Clinic in Minnesota, but it is unclear whether they made the trip.

43. "Annual Report" (Leukemia Society, 1955), Lymphoma and Leukemia Society of America, White Plains, New York.

44. Dargeon, *Cancer in Childhood*, 40.

45. William Dameshek, "Leukemia," *Hygeia* 24/12 (Dec. 1946): 908.

46. Ibid.

47. Ibid., 958.

48. See "Leukemia Studies: Foundation to Award a Prize for Best Paper on the Disease," *New York Times* (9 April 1950): 9, for an announcement for the first prize competition. Later, it ex-

panded its awards contest into a grant-in-aid program designated for outstanding research projects by young investigators and a fellowship source for physicians training in the specialized field of leukemia and related diseases.

49. Ibid.

50. Sidney Farber, Louis K. Diamond, Robert D. Mercer, Robert F. Sylvester, Jr., and James A. Wolff, "Temporary Remissions in Acute Leukemia in Children Produced by Folic Acid Antagonist 4-Aminopteroyl-Glutamic Acid (Aminopterin)," *New England Journal of Medicine* 238 (1948): 787–793.

51. "Hope for Leukemia," *Newsweek* 31 (April 26, 1948): 48.

52. Farber also coordinated exchanges between the pharmaceutical company and his colleague Louis Diamond who was researching anemia. A letter from November 1943 stated, "I made arrangements with Dr. Diamond for trying the 80 per cent extract on suitable cases of anemia. We shall begin when the capsules arrive." Correspondence, Sidney Farber to Dr. Y. Subbarow, 28 Nov. 1943, Dana-Farber Cancer Institute Library, Boston, Massachusetts (hereafter referred to as DFCI).

53. Correspondence, Sidney Farber to Dr. Y. Subbarow, 29 March 1944, DFCI. Unfortunately, this small collection only contains a portion of one side of the correspondence between Farber and Subbarow.

54. Correspondence, Sidney Farber to Dr. Y. Subbarow, 30 June 1944, DFCI. Gaps in correspondence between 1944–1945 and 1945–1948 make it difficult to reconstruct the scientific investigations on diet, vitamins, and disease at Lederle and Boston Children's Hospital during these periods. Later, in a memorial for Subbarow, Farber claimed that his colleague had made aminopterin for Farber at his request.

55. Correspondence, Sidney Farber to Dr. Y. Subbarow, 15 April 1948, DFCI.

56. No written record was kept of this conference.

57. Remarks among Farber's colleagues suggested that this discrepancy raised questions whether Farber was completely truthful when reporting his experimental design, methods, and results.

58. Correspondence, Sidney Farber to Dr. Y. Subbarow, 2 February 1948, DFCI.

59. Wisnia presents a chronological review of the Jimmy Fund's fundraising activities.

60. "The House that Jimmy Built," Program, Sidney Farber, Dedication Ceremony, 7 Jan. 1952, DFCI, 16.

61. Ibid., 8. The previous historical background on the Jimmy Fund is also from speakers participating in the program.

Chapter Three • Death Be Not Proud

1. John Gunther, *Death Be Not Proud: A Memoir* (New York: Harper and Brothers, 1949), 3.

2. Ibid.

3. John Gunther, *A Fragment of Autobiography: The Fun of Writing the Inside Books* (New York: Harper and Row, 1962), 9.

4. Gunther, *A Fragment of Autobiography*, 6.

5. Ibid., 39.

6. Ibid., 60.

7. Ibid.

8. Ibid., 45.

9. John Hayward and Geoffrey Keynes, eds., *The Complete Poetry and Selected Prose of John Donne and the Complete Poetry of William Blake* (New York: Random House, 1941), 239.

10. The book's dust jacket stated that neither the publisher nor the author was profiting from the account and that the proceeds would be donated to cancer research for children. A letter from John to Frances in 1949 concerned the funds. John proposed that they share part of the proceeds from the *Ladies' Home Journal* article and the book. "I thought of dividing the rest," he wrote, "between Deerfield, Memorial, or some other hospital (in the form of a specific endowment in Johnny's name for tumor research, if the sum is considerable enough)." He also considered giving a portion of the money to Gerson or keeping some as reimbursement for their medical expenses. John Gunther to Frances Gunther, [1949], Box 4, Folder 175, Frances Fineman Gunther Papers, 1915–1963, Schlesinger Library, Radcliffe Institute, Harvard University, Cambridge, Massachusetts (hereafter referred to as FGP). Unfortunately, I was unable to determine the final distribution of the profits.

11. L. Emmett Holt to Frances Gunther, 7 June 1929, Carton 2, Folder 46, FGP. Another physician termed Judy's death "an unavoidable accident." Julius A. Miller to Frances Gunther, 28 February 1929, Carton 2, Folder 46, FGP.

12. Frances Gunther, 1 Jan. 1947, Box 1, Folder 15, FGP.

13. Frances Gunther, 8 May 1948, Box 1, Folder 15, FGP.

14. Marginal notes, Draft of "A Word from Frances," n.d., Box 44, Folder 18, John Gunther Collection, Special Collections, Joseph Regenstein Library, University of Chicago, Illinois (hereafter referred to as JGC).

15. Notes labeled "Diary," n.d., Box 44, Folder 18, JGC.

16. Ibid.

17. Gunther, *Death Be Not Proud,* 32.

18. It is unclear what Johnny knew about his illness. After his first surgery, John recorded a brief exchange between Putnam and Johnny in which Putnam stated, "Johnny, what we operated for was a brain tumor" (5). Later in *Death Be Not Proud,* Gunther wrote that Johnny did not know the extent of his condition. He said, "We had to shield him from definite, explicit knowledge" (50). He hid the encyclopedia volume that contained information on his son's tumor. Later, after leaving an x-ray appointment, Johnny asked, "Does this mean I have cancer?" (61). The Gunthers also noted that Johnny's personal writings revealed that he knew more about his tumor than they had realized. Physicians often concealed cancer diagnoses from patients. In some instances, spouses or family members were told the truth about the victim's diagnosis and prognosis so that they could manage the patient's affairs. See Samuel Standard and Helmuth Nathan, *Should the Patient Know the Truth?: A Response of Physicians, Nurses, Clergymen, and Lawyers* (New York: Springer, 1955), for specific sections on children and truth telling. For a description of truth telling practices, see Susan E. Lederer, "Medical Ethics and the Media: Oaths, Codes, and Popular Culture." In Robert G. Baker, Arthur L. Caplan, Linda Emanuel, and Stephen S. Latham, eds., *The American Medical Ethics Revolution* (Baltimore: Johns Hopkins University Press, 1999), 91–103.

19. Gunther, *Death Be Not Proud,* 254.

20. For a comprehensive overview of the field, see two recent collections on child health and

welfare: Alexandra Minna Stern and Howard Markel, eds., *Formative Years: Children's Health in the United States, 1880–2000* (Ann Arbor: University of Michigan Press, 2002), and Roger Cooter, ed., *In the Name of the Child: Health and Welfare, 1880–1940* (London: Routledge, 1992). The first provides a concise overview of the field, including medical specialties, technologies, adolescent health, and changing definitions of childhood diseases. The contributors to the latter volume examined topics such as children's bodies, childhood diseases, child abuse, child labor, and child guidance in comparative articles that crossed geographical boundaries.

21. See Arthur Kleinman, *Illness Narratives: Suffering, Healing, and the Human Condition* (New York: Basic Books, 1988); Howard Brody, *Stories of Sickness* (New Haven, Conn.: Yale University Press, 1987); and Anne Hunsaker Hawkins, *Reconstructing Illness* (West Lafayette, Ind.: Purdue University Press, 1993), for analyses of this literary genre. Other major works on illness narratives include Arthur W. Frank, "The Rhetoric of Self-Change: Illness Experience as Narrative," *Sociology Quarterly* 34 (1993): 39–52; Thomas Couser, *Recovering Bodies: Illness, Disability, and Life Writing* (Madison: University of Wisconsin Press, 1997); and *a/b Auto/Biography Studies: Special Issue: Illness, Disability, and Lifewriting* 6/1 (spring 1991). For a summary of the literature on illness narratives, see Amy Fairchild, "The Polio Narratives: Dialogues with FDR," *Bulletin of the History of Medicine* 75/3 (fall 2001): 491–497.

22. Daniel J. Wilson, *Living with Polio: The Epidemic and Its Survivors* (Chicago: University of Chicago Press, 2005); "A Crippling Fear: Experiencing Polio in the Era of FDR," *Bulletin of the History of Medicine* 72 (1998): 464–495; Daniel J. Wilson, "Covenants of Work and Grace: Themes of Recovery and Redemption in Polio Narratives," *Literature & Medicine* 13 (1994): 22–41; and Fairchild, "The Polio Narratives," 491.

23. Kathryn Black, *In the Shadow of Polio: A Personal and Social History* (Reading, Mass.: Addison-Wesley, 1996). One boy referred to his room as a "way station for death" and then recounted how the patients knew that someone had died on the ward. "For those in iron lungs," he recalled, "the death signal was the passing of a nurse down the row of tanks, turning each mirror so the patients couldn't watch while a body was wheeled out in a now silent machine" (64).

24. Gunther, *Death Be Not Proud,* 26.

25. Ibid., 41.

26. Ibid., 42.

27. Ibid., 54. The pathological diagnosis was also recorded in the Medical Record, Roentgentherapy Sheet, Box 44, Folder 7, JGC.

28. Related work on the postwar dynamic includes Angela N. H. Creager, "Tracing the Politics of Changing Postwar Research Practices: The Export of 'American' Radioisotopes to European Biologists," *Studies in History and Philosophy of Science. Part C: Biological and Biomedical Sciences* 33 (2002): 367–388; and Stuart M. Feffer, "Atoms, Cancer, and Politics: Supporting Atomic Science at the University of Chicago, 1944–1950," *Historical Studies in the Physical and Biological Sciences* 22 (1992): 233–261. Note that John Gunther later wrote to researchers at the school looking for therapeutic alternatives.

29. Medical record, Box 44, Folder 7, JGC.

30. Gunther, *Death Be Not Proud,* 57.

31. Ibid., 58.

32. Lester A. Mount, Medical Report, Examination of John Gunther, 20 July 1946, Box 44, Folder 7, JGC.

33. Clement B. Masson to Cornelius Traeger, 16 July 1946, Box 44, Folder 7, JGC. Masson replaced Putnam after he moved to California.

34. Gunther, *Death Be Not Proud,* 74.

35. Ibid. Also see Report of Dr. Penfield, 27 July 1946, Box 44, Folder 8, JGC.

36. Gunther, *Death Be Not Proud,* 71.

37. Ibid., 76.

38. John Gunther to Ernest O. Lawrence, 30 July 1946, Box 44, Folder 8, JGC. The identical letters sent to Lawrence, Evans, and Hutchins were dated 30 July 1946. Notes from John Gunther's archival collection suggest that he may have contacted at least a dozen other scientists and experts over the course of his son's illness.

39. Ibid.

40. Ibid.

41. Gunther, *Death Be Not Proud,* 81–82.

42. Ibid., 32.

43. Ibid., 171–172.

44. Vannevar Bush, *Science: The Endless Frontier: A Report to the President on a Program for Postwar Scientific Research* (Washington, D.C.: Government Printing Office, 1945).

45. Paul Starr, *The Social Transformation of American Medicine: The Rise of a Sovereign Profession and the Making of a Vast Industry* (New York: Basic Books, 1982), 335–336. See Allan M. Brandt and Martha Gardner, "The Golden Age of Medicine." In Roger Cooter and John Pickstone, eds., *The History of Twentieth-Century Medicine* (Amsterdam: Harwood, 2000), 21–37.

46. John C. Burnham, "American Medicine's Golden Age: What Happened to It?" *Science* 215 (19 March 182): 1474–1479.

47. Gunther, *Death Be Not Proud,* 89.

48. Max Gerson, *A Cancer Therapy: Results of Fifty Cases* (New York: Whittier Books, 1958).

49. Gunther, *Death Be Not Proud,* 117.

50. Max Gerson to John and Frances Gunther, 28 Dec. 1946, Box 44, Folder 8, JGC.

51. Tracy Putnam to John Gunther, 15 Jan. 1947, Box 44, Folder 8, JGC.

52. The Gunthers' use of Gerson's diet therapy both supports and contradicts the conclusions of David M. Eisenberg, Ronald C. Kessler, Cindy Foster, Frances E. Norlock, David R. Calkins, and Thomas L. Delbanco, "Unconventional Medicine in the United States—Prevalence, Costs, and Patterns of Use," *New England Journal of Medicine* 328/4 (28 Jan. 1993): 246–252. In this study, Eisenberg and his colleagues found that nonblack persons 25–49 years of age with advanced education and higher incomes were the most common users of alternative therapies. Most users, however, sought unconventional therapies for chronic, not life-threatening conditions. They also commonly enlisted the aid of both unconventional healers and medical doctors simultaneously, but did not disclose this information to their physician.

53. Gunther, *Death Be Not Proud,* 144.

54. Ibid., 151.

55. Ibid., 168.

56. Ibid., 189.

57. New York City newspapers such as the *New York Times, New York Herald Tribune, New York Sun,* and *New York World Telegram* published Johnny's obituary on 1 and 2 July 1947. It was

also picked up by the Associated Press and printed in over forty newspapers from San Francisco, California to Des Moines, Iowa.

58. Gunther, *Death Be Not Proud,* 194.

59. Finding guide, JGC. John Gunther, "Death Be Not Proud," *Ladies' Home Journal* (February 1949): 38–39, 87–116. John Gunther, "Death Be Not Proud," *Reader's Digest* (March 1949): 129–146.

60. The finding aid for the section of the John Gunther Papers devoted to *Death Be Not Proud* states, "The correspondence is the most extensive part of the material on this work. There are two folders of personal correspondence, all post-publication, from friends of Gunther commenting on the book. There are also four boxes, containing approximately four thousand letters from the general public" (Finding aid, JGC).

61. Walter Duranty, *New York Herald Tribune Weekly Book Review* (6 February 1949). Other reviews included: Lewis Gannett, *New York Tribune* (3 February 1949); [Review], *New Yorker* (26 February 1949): 98; Pamela Taylor, "Johnny Gunther's Gallant Battle," *Saturday Review of Books* (5 March 1949): 27; M.C. Scoggin, "Outlook Tower," *Horn Books* 25 (May 1949): 241; and [Review], *Time* 53 (7 February 1949): 92, 94.

62. Nathaniel M. Guptill to John Gunther, 5 April 1949, Box 46, Folder 2, JGC.

63. E. Burdette Backus, 5 April 1953, Box 45, Folder 4, JGC.

64. Niels Nelson to John Gunther, n.d., Box 47, Folder 5, JGC. Niels Nielsen, "More Than Any Sermon," *Ladies' Home Journal* (April 1949): 6.

65. Paul B. Hoeber to John Gunther, 17 May 1949, Box 48, Folder 11, JGC.

66. Walter L. Palmer to John Gunther, 9 February 1949, Box 44, Folder 23, JGC.

67. Ibid.

68. Charles Huggins to John Gunther, 7 July 1952, Box 48, Folder 11, JGC.

69. Yvette Fay Francis to John Gunther, 24 Sept. 1949, Box 45, Folder 10, JGC.

70. Ibid.

71. Mary L. Davis to John Gunther, 25 July 1962, Box 33, Folder "Fan Mail," JGC.

72. Jean Kuffner to John Gunther, 7 March 1949, Box 46, Folder 8, JGC.

73. Lois Payne to John Gunther, 5 March 1949, Box, 47, Folder 6, JGC.

74. Mrs. D. B. Hopkins to John Gunther, n.d., Box 46, Folder 5, JGC.

75. Betty Jane Sheffler to John Gunther, 8 March 1949, Box 48, Folder 1, JGC.

76. Mary Reilly to John Gunther, 22 April 1958, Box 47, Folder 10, JGC.

77. Beverly Goodell to John Gunther, 5 Nov. 1958, Box 46, Folder 2, JGC.

78. Mary A. Dotterer to John and Frances Gunther, 10 March 1949, Box 45, Folder 7, JGC.

79. [Unreadable] Baasch to John and Frances Gunther, Box 45, Folder 3, JGC.

80. Gunther, *Death Be Not Proud,* 251.

81. Jane Raborg to John Gunther, 15 February 1949, Box, 47, Folder 8, JGC.

82. Marianne Peters to Frances Gunther, 13 February 1949, Carton 2, Folder 58, FGP.

83. Elizabeth V. Guthrie to John and Frances Gunther, 18 Dec. 1949, Box 46, Folder 1, JGC.

84. Ibid.

85. Elizabeth V. Guthrie to John and Frances Gunther, 25 June 1951, Box 46, Folder 1, JGC.

86. Elizabeth V. Guthrie to John and Frances Gunther, 31 July 1951, Box 46, Folder 1, JGC.

87. Mrs. Martin H. Byrd to John Gunther, n.d., Box 45, Folder 2, JGC.

88. Mrs. T. H. R. to John Gunther, 23 February 1949, Box 47, Folder 9, JGC.

89. See Robert N. Proctor, *Cancer Wars: What We Know and Don't Know about Cancer* (New York: Basic Books, 1995) for an examination of the major theories of cancer causation (behavioral, environmental, and hereditary) within the larger political history of the disease. Proctor analyzes the way in which various groups promoted or suppressed particular causes in their pursuit of research funding, consumer support, or political power.

90. Frustrated parents asked why physicians and scientists did not know more about the origins of tumors or about their swift growth.

91. Mrs. Lewis Orrell, Jr., to John Gunther (n.d), Box 48, Folder 11, JGC.

92. Ibid.

93. Mrs. Cramer to Frances Gunther, February 1949, Carton 2, Folder 58, FGP.

94. Mrs. Raymond Kaplan to Frances Gunther, 7 February 1949, Carton 2, Folder 58, FGP.

95. Ibid.

96. J. Davis to John Gunther, 25 March 1949, JGC.

97. Limited archival documentation suggests that Gunther neither explicitly endorsed Max Gerson's healing method nor sent contact information to readers.

98. Jean D. Roessler to John Gunther, 18 February 1949, Box 47, Folder 10, JGC.

99. Genevieve Christianson to John and Frances Gunther, 13 March 1950, Box 45, Folder 6, JGC.

100. Ibid.

101. Viviana Zelizer, *Pricing the Priceless Child: The Changing Social Value of Children* (New York: Basic Books, 1985); Elaine Tyler May, *Homeward Bound: American Families in the Cold War Era* (New York: Basic Books, 1988), xiv; and Susan Hartmann, *The Homefront and Beyond: American Women in the 1940s* (Boston: Twayne, 1982).

102. Landon Jones, *Great Expectations: America and the Baby Boom Generation* (New York: Coward, McCann and Geoghegan, 1980).

103. Wendy Kozol, *Life's America: Family and Nation in Postwar Photojournalism* (Philadelphia: Temple University Press, 1994).

104. Hieda N. Janovak to John Gunther, 22 Jan. 1959, Box 46, Folder 6, JGC.

105. William Rodgers to John Gunther, 24 Jan. 1949, Box 47, Folder 9, JGC.

106. Sociologists have mined literary sources to examine how expressions of maternal grief over infant death changed over time. They found that only one magazine, *True Stories,* a publication for working-class women, featured stories of maternal grief in the mid-twentieth century. Wendy Simonds and Barbara Katz Rothman, *Centuries of Solace: Expressions of Maternal Grief in Popular Literature* (Philadelphia: Temple University Press, 1992).

107. Mrs. A. J. Hummel to John Gunther, 22 March 1949, Box 46, Folder 5, JGC.

108. Loretta Maxwell to John Gunther, 23 Aug. 1952, Box 46, Folder 7, JGC.

109. Helen L. Kaufmann to John Gunther, 7 March 1949, Box 46, Folder 8, JGC.

110. Margaret J. Oberfelder to John Gunther, 1 February 1949, Box 47, Folder 6, JGC.

111. Susan Sontag, *Illness as Metaphor* (New York: Farrar, Straus and Giroux, 1978). Sontag wrote that such metaphors "are responses to a disease thought to be intractable and capricious—that is, a disease not understood—in an era in which medicine's central premise is that all diseases can be cured. Such a disease is, by definition, mysterious" (5). Sontag carefully traced

the changing, varied uses of tuberculosis and cancer metaphors but ultimately derided their use as simplistic.

112. Mrs. G. Clifford to John Gunther, 16 March 1949, Box 45, Folder 5, JGC.

113. Lucy P. Gregg to John Gunther, 1 February 1949, Box 46, Folder 1, JGC.

114. Ibid.

115. Chuck Gordon to John and Frances Gunther, 14 March 1949, Box 46, Folder 2, JGC.

116. Gertrude Hepworth to John and Frances Gunther, Box 46, Folder 5, JGC.

117. Mildred Mize to John Gunther, 1 February 1949, Box 47, Folder 4, JGC.

118. After first appearing in hardcover in 1949, later editions of the book were published in 1953, 1966, 1971, 1992, and 1998. In 1975, *Death Be Not Proud* was aired as a television movie. More recently, lines from *Death Be Not Proud* reappeared in *Wit,* a Pulitzer Prize–winning play adapted into a television movie. The play's protagonist, Vivian Bearing, a British professor and scholar of Donne's Holy Sonnets, reflected upon her analysis of *Death Be Not Proud* when faced with her own imminent death from metastatic ovarian cancer. Through performances of *Wit* and the slogan "Death Be Not Ovarian Cancer," cancer organizations have used the play to promote gynecological cancer screenings and to raise cancer awareness in women.

119. Radio script, North Carolina Division of the American Cancer Society, Box 48, Folder 13, JGC.

120. Russell M. Viner, and Janet Golden, "Children's Experiences of Illness." In Roger Cooter and John Pickstone, eds., *The History of Twentieth-Century Medicine* (Amsterdam: Harwood, 2000), 576.

Chapter Four • "Against All Odds"

1. Angela Burns to John Gunther, 29 Jan. 1949, Box 45, Folder 2, John Gunther Collection, Special Collections, Joseph Regenstein Library, University of Chicago, Illinois.

2. Ibid.

3. Allan M. Brandt and Martha Gardner, "The Golden Age of Medicine." In Roger Cooter and John Pickstone, eds., *The History of Twentieth Century Medicine* (Amsterdam: Harwood, 2000), 21–37.

4. Emil J. Freireich and Noreen A. Lemak, *Milestones in Leukemia Research and Therapy* (Baltimore: Johns Hopkins University Press, 1991), and John Laszlo, *The Cure of Childhood Leukemia: Into the Age of Miracles* (New Brunswick, N.J.: Rutgers University Press, 1995). Freireich, Lemak, and Laszlo have written historical accounts that characterized the field of acute leukemia research in the 1950s and 1960s as a history of scientific progress or "medical milestones" that led to temporary remissions in children suffering from the disease and treatments for other types of cancer. Other physicians, primarily from the National Cancer Institute, have also recorded this series of events as they related to institutional achievements and challenges. C. Gordon Zubrod, "The Cure of Cancer by Chemotherapy: Reflections on How It Happened," *Medical and Pediatric Oncology* 8/2 (1980): 107–114, and Emil Frei III, "Intramural Therapeutic Research at the National Cancer Institute, Department of Medicine: 1955–1965," *Cancer Treatment Reports* 68/1 (Jan. 1984): 21–30, are only a few examples of the retrospective accounts.

5. See Chris Feudtner, *Bittersweet: Diabetes, Insulin, and the Transformation of Illness* (Chapel Hill: University of North Carolina Press, 2003), for another "patient-centered" disease history. Feudtner found that the revolution of insulin therapy during 1921–1922 not only transformed diabetes from an acute to a chronic disease but also caused short- and long-term complications that impacted doctors, nurses, and the family and patient.

6. Institute of Medicine, *Veterans at Risk: The Health Effects of Mustard Gas and Lewisite* (Washington, D.C.: Institute of Medicine, 1993), 44–45.

7. James Phinney Baxter III, *Scientists against Time* (Boston: Little, Brown, 1946), 266. See part 3—Chemistry and War, chapter 18, "Why Not Gas?" for the most detailed section on this topic. For a discussion of chemical development and the important intersections between national defense and insect control from World War I to the post–Cold War era, see Edmund Russell, *War and Nature: Fighting Humans and Insects with Chemicals from World War I to* Silent Spring (Cambridge: Cambridge University Press, 2001).

8. E. B. Krumbhaar, "Role of the Blood and the Bone Marrow in Certain Forms of Gas Poisoning. I. Peripheral Blood Changes and Their Significance," *Journal of the American Medical Association* 72 (1919): 39–41, and E. B. Krumbhaar and H. D. Krumbaar, "The Blood and Bone Marrow in Yellow Cross Gas (Mustard Gas) Poisoning: Changes Produced in the Bone Marrow of Fatal Cases," *Journal of Medical Research* 40 (1919): 497–558.

9. Louis S. Goodman, Maxwell W. Wintrobe, William Dameshek, Morton J. Goodman, Alfred Gilman, and Margaret T. McLennan "Nitrogen Mustard Therapy: Use of Methyl-Bis(Beta-Chloroethyl)amine Hydrochloride and Tris(Beta-Chloroethyl)amine Hydrochloride for Hodgkin's Disease, Lymphosarcoma, Leukemia and Certain Allied and Miscellaneous Disorders," *Journal of the American Medical Association* (21 Sept. 1946): 126. Although the initial clinical trial was held in May 1942, the results were held until 1946 because of wartime secrecy. Facets of the work were published as A. Gilman, "Symposium on Advances in Pharmacology Resulting from War Research: Therapeutic Applications of Chemical Warfare Agents," *Federation Proceedings* 5 (1946): 285–292; A. Gilman and F. S. Philips, "The Biological Actions and Therapeutic Application of the b-Choroethylamines and Sulfides," *Science* 103 (1946): 409–415; C. P. Rhoads, "Work of Chemical Warfare Service, esp. on Nitrogen Mustards," *Journal of Mt. Sinai Hospital* 13 (1947): 299–309, and "Nitrogen Mustards in the Treatment of Neoplastic Disease: Official Statement," *Journal of the American Medical Association* 131 (1946): 656–658; and A. Gilman and M. Cattell, "Systemic Agents: Action and Treatment." In E. C. Andrus, D. W. Bronk, G. A. Carden, Jr., C. S. Keefer, J. S. Lockwood, J. T. Wearn, and M. C. Winternitz, *Advances in Military Medicine: Science in World War II: Office of Scientific Research and Development* (Boston: Little, Brown, 1948), 546–564.

10. Goodman et al., 131.

11. Rhoads, "Nitrogen Mustards in the Treatment of Neoplastic Disease."

12. S. Farber, Louis K. Diamond, Robert D. Mercer, Robert F. Sylvester, Jr., and James A. Wolff, "Temporary Remissions in Acute Leukemia in Children Produced by Folic Acid Antagonist 4-Aminopteroyl-Glutamic Acid (Aminopterin)," *New England Journal of Medicine* 238 (1948): 787. Aminopterin was modified into an equally effective but less toxic form called amethoperin (methotrexate). Like Farber, M. C. Li at the National Cancer Center stimulated disbelief and curiosity after utilizing methotrexate to cure a woman with choriocarcinoma, a rare reproductive cancer.

13. Angela Burns to John Gunther, 29 Jan. 1949, Box 45, Folder 2, JGC.

14. Ibid.

15. "Leukemia in Children and Adults," *CA: A Bulletin of Cancer Progress* 7/2 (March 1957): 54. Photos were supplied courtesy of J. H. Burchenal, R. R. Ellison, M. L. Murphy, and T. C. Tan, Division of Clinical Chemotherapy, Sloan-Kettering Institute, and Departments of Medicine and Pediatrics, Memorial Center for Cancer and Allied Diseases.

16. Angela Burns to John Gunther, 29 Jan. 1949, Box 45, Folder 2, JGC.

17. Ibid.

18. Ibid.

19. "The House That Jimmy Built," Program, Sidney Farber, Dedication Ceremonies (7 Jan. 1952). In George E. Foley, *The Children's Cancer Research Foundation: The House That "Jimmy" Built, The First Quarter-Century* (privately published, held at Dana-Farber Research Institute Library), 17. Farber later articulated an expanded definition in Farber, "Management of the Acute Leukemia Patient and Family," *CA: Cancer Journal for Clinicians* 15/1 (Jan.–Feb. 1965): 14–17.

20. See Richard Cabot, *Social Service and the Art of Healing* (New York: Moffat, Yard, 1909), for a description of his concept.

21. Richard Cabot, *Social Work: Essays on the Meeting Ground of Doctor and Social Worker* (Boston: Houghton Mifflin, 1919), vii, xxv.

22. See Janet Golden, ed., *Infant Asylums and Children's Hospitals: Medical Dilemmas and Developments, 1850–1920* (New York: Garland, 1989) for a collection of relevant primary sources. For a case study, see Helen Hughes Evans, "Hospital Waifs: The Hospital Care of Children in Boston, 1860–1920" (Ph.D. diss., Harvard University, 1995).

23. Foley, *The Children's Cancer Research Foundation*, 17.

24. Ibid.

25. In *Strangers at the Bedside: A History of How Law and Bioethics Transformed Medical Decision Making* (New York: Basic Books, 1991), David Rothman argues that the urgency of wartime research continued into the postwar decades, encouraging the physician investigator to blur the lines between research subject and patient.

26. "Tower of Hope," *Reader's Digest* (Sept. 1949): 13. This article was condensed from the 27 June 1949 issue of *Time.*

27. "Tower of Hope," *Reader's Digest.*

28. Mary A. MacRostie, "What Nurses Are Doing for Children with Leukemia," *R.N.: A Journal for Nurses* 20/10 (Oct. 1957): 50.

29. Mary Stone Brodish, "The Nurse's Role in the Care of Children with Acute Leukemia," *American Journal of Nursing* 58/11 (Nov. 1958): 1573. Brodish wrote the article after periods of observation in the pediatric outpatient clinic of the Grace–New Haven Hospital, the homes of several families being followed by the clinic, Memorial Hospital, and the Jimmy Fund Building.

30. Ibid., 1574.

31. Ibid.

32. See James Robertson, *Young Children in Hospitals* (New York: Basic Books, 1958) for a criticism of British care.

33. Josephine Sever, *Johnny Goes to the Hospital* (Boston: Houghton Mifflin, 1953). See also Nancy Dudley, *Linda Goes to the Hospital* (New York: Coward-McCann, 1953).

34. See David W. Adams, *Childhood Malignancy: The Psychosocial Care of the Child and His Family* (Springfield, Ill.: Charles C Thomas, 1979), 4–16, for a comprehensive review of psychological studies of pediatric cancer sufferers and their families beginning in 1950.

35. Mary F. Bozeman, Charles E. Orbach, and Arthur M. Sutherland, "Psychological Impact of Cancer and Its Treatment. III. The Adaptation of Mothers to the Threatened Loss of Their Children through Leukemia. Part I," *Cancer* 8/1 (Jan.–Feb. 1955): 1.

36. Beatrix Cobb, "Psychological Impact of Long Illness and Death of a Child on the Family Circle," *Journal of Pediatrics* 49 (1956): 746–751.

37. Ibid, 746.

38. Julius B. Richmond and Harry A. Waisman, "Psychological Aspects of Management of Children with Malignant Diseases," *American Journal of Diseases of Children* 89 (1955): 42–47.

39. Ibid., 45.

40. Harry M. Marks, "Cortisone, 1949: A Year in the Political Life of a Drug," *Bulletin of the History of Medicine* 66 (1992): 419–439. Marks documented the efforts to ration the limited but coveted cortisone supply.

41. S. Farber, V. Downing, H. Shwachman, R. Toch, R. Appleton, F. Heald, J. P. King, and D. Feriozi, "The Effect of ACTH in Acute Leukemia in Childhood." In J. R. Mote, ed., *Proceedings of the First Clinical ACTH Conference* (Philadelphia: Blakiston, 1950), 226–234.

42. S. Farber, H. Schwachman, R. Toch, V. Downing, B. H. Kennedy, and J. Hyde, "The Action of ACTH and Cortisone in Acute Leukemia." In J. R. Mote, ed. *Proceedings of the Second Clinical ACTH Conference* (New York: Blakiston, 1951), 251.

43. Ibid., 287.

44. "Cortisone and ACTH Prolong Leukemia Victims' Lives," *Science News Letter* 61/13 (29 March 1952): 200. See Elizabeth M. Kingsley Pillers, Joseph H. Burchenal, Leonard P. Eliel, and Olaf H. Pearson, "Resistance to Corticotropin, Cortisone, and Folic Acid Antagonists in Leukemia," *Journal of the American Medical Association* (22 March 1952): 987–994.

45. Newsletter, *CA: A Bulletin of Cancer Progress* 4/2 (March 1954).

46. "2 Leukemia Victims Freed at Bellevue," *New York Times* (22 March 1950): 6.

47. Ibid.

48. Ibid.

49. Ibid.

50. "Leukemia Poster Girl Dies," *New York Times* (7 May 1950): 96.

51. For a report of a limited clinical trial of 6-MP in advanced acute leukemia in children, see Joseph H. Burchenal, David Karnofsky, M. Lois Murphy, Ruth Ellison, and C. P. Rhoads, "Effects of 6-Mercaptopurine in Man," *Proceedings of the American Association of Cancer Research* 1 (1953): 7. See J. H. Burchenal, M. L. Murphy, R. R. Ellison, M. P. Sykes, T. C. Tan, L. A. Leone, D. A Karnofsky, L. F. Craver, H. W. Dargcon, and C. P. Rhodes, "Clinical Evaluation of a New Antimetabolite, 6-Mercaptopurine, in the Treatment of Leukemia and Allied Diseases," *Blood* 8 (1953): 965–999, for a set of conference papers on the toxic effects, mechanism, action on tumors, and clinical experience with 6-MP on leukemia and other cancers at varied research institutions.

52. Hitchings and Elion shared the Nobel Prize in Medicine or Physiology in 1988 for their development of antimetabolite compounds and their elucidation of the basic principles of DNA synthesis inhibition.

53. J. H. Burchenal, D. A. Karnofsky, M. L. Murphy, R. R. Ellison, M. P. Sykes, C. T. Tan, A. C. Mermann, M. Yuceoglu, and C. P. Rhoads, "Clinical Evaluation of 6-Mercaptopurine in the Treatment of Leukemia," *American Journal of Medical Science* 228 (Oct. 1954): 372. On 30 April and 1 May 1954, a 6-MP conference was held at the New York Academy of Sciences and co-chaired by George H. Hitchings and C. P. Rhoads. The papers were published in the *Annals of the New York Academy of Sciences* 60/2 (6 Dec. 1954): 183–508.

54. Burchenal, 377.

55. Ibid.

56. "Toward Chemical Cures for Cancer," *New York Times* (11 July 1954): 9.

57. "Temporary Leukemic Relief," *Science News Letter* 65/1 (2 Jan. 1954): 5.

58. C. Gordon Zubrod, "History of the Cancer Chemotherapy Program," *Cancer Chemotherapy Reports* 50/7 (Oct. 1966): 351. For a history of the application of industrial research methods to screening programs at Sloan-Kettering Institute and the Institute for Cancer Research in Philadelphia, see R. F. Bud, "Strategy in American Cancer Research after World War II: A Case Study," *Social Studies of Science* 8/4 (Nov. 1978): 425–459.

59. Peter Keating and Alberto Cambrosio, "From Screening to Clinical Research: The Cure of Leukemia and the Early Development of the Cooperative Oncology Groups, 1955–1966," *Bulletin of the History of Medicine* 76 (2002): 299–334. Keating and Cambrosio trace the events of this eleven-year period—from the establishment of a new screening system to the 1966 reorganization under C. Gordon Zubrod that created a fissure between the two activities. See also Ilana Löwy, *Between Bench and Bedside: Science, Healing, and Interleukin-2 in a Cancer Ward* (Cambridge, Mass.: Harvard University Press, 1996), 42–59.

60. Zubrod, 353.

61. Shimkin wrote, "A stimulating factor, as so often is the case, is that a neighborhood child of a staff member of an influential Congressman had died of leukemia." Michael B. Shimkin, "As Memory Serves: An Informal History of the National Cancer Institute, 1937–57," *Journal of the National Cancer Institute* 59/2 (Aug. 1977): 591.

62. "Cancer Cure Search: 'Crash' Program to Win Fight against Cancer Inaugurated under Direction of a Committee Combining Government, Public and Industry Representatives," *Science News Letter* (11 June 1955), 371.

63. "Cancer Cure Search," 371.

64. "Humans Test Cancer Drugs," *Science News Letter* (18 April 1959): 246.

65. Kenneth Endicott, "The Chemotherapy Program," *Journal of the National Cancer Institute* 19/2 (Aug. 1957): 275–277.

66. Criticism of the massive empirical screening program and the rodent tumor system abounded. Murray Shear, a researcher in the Laboratory of Chemical Pharmacology at the National Cancer Center, wrote, "Time and again these 'pure' science colleagues of ours have warned us that we are wasting our own time and our institutions' money by fumbling empirically in the dark, without rational bases for our chemotherapy experiments." Other critics questioned the merit of testing possible agents in mice, rabbits, and dogs, saying that their activity in animals was unrelated to their benefits to humans. Murray J. Shear, "Role of the Chemotherapy Research Laboratory in Clinical Cancer Research," *Journal of the National Cancer Institute* 12 (1951–1952): 574.

67. In 1953, the National Cancer Center had established a clinical center for cancer patients and James Holland initiated a clinical leukemia program at the institute.

68. K. M. Endicott, "The National Chemotherapy Program," *Journal of Chronic Diseases* 8 (1958): 171–178. In the late 1940s and 1950s physicians compiled information about the childhood cancers treated within their own institution or state. Physicians attempted to identify trends within their limited samples, but noted that the small number of cases, the variety of departments treating cancer, and the diversity of cases presented hindered their efforts to comprehensively, accurately analyze childhood cancers. Frustrated physicians called for a "pooling of experience" to gain a "composite experience." See Dorothy H. Andersen, "Tumors of Infancy and Childhood. I. A Survey of Those Seen in the Pathology Laboratory of the Babies Hospital during the Years 1935–1950," *Cancer* 4/4 (July 1951): 890–906; Ralph E. Knutti, "Malignant Tumors of Childhood," *California Medicine* 76/1(Jan. 1952): 27–29; Vincent H. Handy, "The Occurrence of Malignancies in Children," *New York State Medical Journal* 56/2 (15 Jan. 1956): 258–260; and James B. Arey, "Cancer in Infancy and Childhood," *Pennsylvania Medical Journal* 55/6 (June 1952): 553–557.

69. Researchers at Memorial under Burchenal initially resisted strict biometric controls in their studies and those affiliated with Farber rejected the use of controls. Emil Frei III, "Intramural Therapeutic Research at the National Cancer Institute, Department of Medicine: 1955–1965," *Cancer Treatment Reports* 68/1 (Jan. 1984): 21. See Hugh L. Davis, John R. Durant, and James F. Holland, "Interrelationships: The Groups, the NCI, and Other Governmental Agencies." In Barth Hoogstraten, ed., *Cancer Research: Impact of the Cooperative Groups* (New York: Masson, 1980), 371–390. For a history of clinical trials and the relationship between rigorous experimentation and statistics in the late twentieth century, see Harry M. Marks, *The Progress of Experiment: Science and Therapeutic Reform in the United States, 1900–1990* (Cambridge: Cambridge University Press, 1997).

70. "NIH Center Applies Team Approach to Research in Cancer of Children," *Scope Weekly* 3/1 (1 Jan. 1958): 1—a publication of the Upjohn Company by Physicians News Service.

71. Ibid.

72. Ibid., 6.

73. Ibid.

74. Henry F. Bisel, "Clinical Aspects of the Cancer Chemotherapy Program," *Current Research in Cancer Chemotherapy* 5 (1956): 7. See also Henry F. Bisel, "Criteria for the Evaluation of Response to Treatment in Acute Leukemia," *Blood* 11/7 (1956): 676–677.

75. Angela Burns to John Gunther, 29 Jan. 1949, Box 45, Folder 2, JGC.

76. Ibid.

77. Ibid.

78. Berger, Meyer, "About New York: Skipper Jackson's Story Disproves the Legend That City People Are Cold and Hard," *New York Times* (17 June 1955): 25.

79. "Leukemia Victim, 3, Succumbs in Sleep," *New York Times* (23 June 1955): 15.

80. Jonathan Engel, *Doctors and Reformers: Discussion and Debate Over Health Policy, 1925–1950* (Columbia: University of South Carolina Press, 2002); Daniel M. Fox, *Health Policies, Health Politics: The British and American Experience, 1911–1965* (Princeton, N.J.: Princeton University Press, 1986); Ronald L. Numbers, "The Specter of Socialized Medicine: American Physicians and Compulsory Health Insurance." In Ronald L. Numbers, ed., *Compulsory Health Insurance: The Continuing American Debate* (Westport, Conn.: Greenwood Press, 1982), 3–24.

81. "Chemotherapy of Leukemia," *CA: A Bulletin of Cancer Progress* 5/5 (Sept. 1955): 154.

82. Barbara Bush, *Barbara Bush: A Memoir* (New York: Scribner and Sons, 1994), 42–52.

83. Ibid, 43.

84. Ibid., 47. The idea that childhood cancers were contagious was expressed in few narratives or other publications from the time, but it is clear from this quote that this isolation or perceived stigmatization was a painful part of the Bush family's experience.

85. Ibid., 48

86. See Nancy Tomes, *The Gospel of Germs: Men, Women, and the Microbe in American Life* (Cambridge, Mass.: Harvard University Press, 1998), and Eileen Margerum, "The Child in American Advertising, 1890–1960: Reflections of a Changing Society." In Harry Eiss, ed., *Images of the Child* (Bowling Green, Ohio: Bowling Green State University Popular Press, 1994), 337, for analysis of how companies incorporated depictions of children in their advertisements to promote a wide array of consumer goods including foods, personal hygiene products, and household cleaning supplies. Anne Higonnet, *Pictures of Innocence: The History and Crisis of Ideal Childhood* (New York: Thames and Hudson, 1998), 116, discusses how photographers captured realistic images of the "child in peril" on film and persuade middle-class viewers to support health and welfare reforms.

87. For examples of the National Foundation for Infantile Paralysis's posters and a critical discussion of their use of children as a fundraising tool, see Jane S. Smith, *Patenting the Sun: Polio and the Salk Vaccine* (New York: William Morrow, 1990), 82–83. Smith, wrote, "Cute little kids on crutches, kids from your hometown, were what opened the wallets and the coin purses" (83). For an extended analysis of poster children and the myths told through promotional images, the complex goals of the modern hospitals and health organizations, and the exploitation of young patients through cancer advocacy and marketing, see Gretchen Krueger, "'For Jimmy and the Boys and Girls of America': Publicizing Childhood Cancers in Twentieth Century America," *Bulletin of the History of Medicine: Special Issue: Cancer in the Twentieth Century* 8/1 (spring 2007): 70–93.

88. There is an extensive body of scholarship on various aspects of polio, including research, treatment, popular responses to polio, the experiences of polio patients, and the ongoing struggle of post-polio patients. See John R. Paul, *A History of Poliomyelitis* (New Haven, Conn.: Yale University Press, 1971); Jane S. Smith, *Patenting the Sun: Polio and the Salk Vaccine* (New York: William Morrow, 1990); Naomi Rogers, *Dirt and Disease: Polio before FDR* (New Brunswick, N.J.: Rutgers University Press, 1996); and Tony Gould, *A Summer Plague: Polio and Its Survivors* (New Haven, Conn.: Yale University Press, 1995).

89. Smith, 35.

90. "The President's Birthday Address," *New York Times* (30 Jan. 1944): 33.

91. Tony Gould, *A Summer Plague: Polio and Its Survivors* (New Haven, Conn.: Yale University Press, 1995), 112.

92. Susan Lederer, "Orphans as Guinea Pigs: American Children and Medical Experimenters, 1890–1930." In Roger Cooter, ed., *In the Name of the Child: Health and Welfare: 1880–1940* (London: Routledge, 1992), 96–123. Lederer argued that from the late nineteenth century until the 1930s, physicians used children as research subjects more frequently than claimed in medical journal reviews. By drawing from two case studies involving diagnostic tests for tuberculosis and syphilis, she demonstrated the close link between antivivisectionist protests and the views regarding the submission of children to research studies. For an extended discussion of

experimentation using child subjects, see Susan Lederer, *Subjected to Science: Human Experimentation in America before the Second World War* (Baltimore: Johns Hopkins University Press, 1995).

93. "A Story-Editorial: Jennifer and the Sword," *ACS Bulletin* 2/14 (6 April 1953): 1.

94. "National Meeting Spurs Enthusiasm for April Crusade," *ACS Bulletin* 4/11 (28 February 1955): 1.

95. "Prize Story Depicts Fight for Survival: Quick Action Meant Life to Doomed Boy," *ACS Bulletin* 2/23 (13 July 1953): 4.

96. "White House Ceremony Opens Cancer Crusade," *ACS Bulletin* 4/16 (11 April 1955): 1.

97. "Leroy Curtis on New York Visit Wins Attention for Crusade," *ACS Bulletin* 4/17 (18 April 1955): 4.

98. "Ed Sullivan Has Whirlwind Day at Cincinnati," *ACS Bulletin* 5/9 (5 March 1956): 3.

99. "Dramatic Radio Program Features Parents of Young Leukemia Patient," *ACS Bulletin* 3/18 (3 May 1954): 1.

100. J. Robert Moskin, "Cancer the Child Killer," *Limelight* (published by *Look*) 2/6 (1955): 2.

101. Ibid., 3–4.

102. Ibid.

103. Ibid., 7.

104. Angela Burns to John Gunther, 29 Jan. 1949, Box 45, Folder 2, JGC.

105. For a comparative analysis of the modern decision between experimental therapy and palliative care for an advanced, incurable cancer, see Ilana Löwy, "'Nothing More to Be Done': Palliative Care versus Experimental Therapy in Advanced Cancer," *Science in Context* 8 (1995): 209–229.

106. "Million-to-One Blow," *New York Times* (25 February 1956): 21.

107. "Twin Dies of Leukemia," *New York Times* (27 February 1956): 21, and "2d Twin Dies of Leukemia," *New York Times* (11 April 1956): 31.

108. This figure was given in "Convict Joins Own Blood Stream to That of Girl Dying of Cancer," *New York Times* (4 June 1949): 1. Other sources estimated that as many as thirty-six pints of blood had been shared.

109. "Convict's Blood Gift Fails to Save Girl," *New York Times* (15 June 1949): 31.

110. "Life from a Lifer," *Time* 53 (13 June 1949): 65.

111. "Terrible Transfusion," *Newsweek* 33 (13 June 1949): 49.

112. Jon M. Harkness, "Research Behind Bars: History of Nontherapeutic Medical Research" (Ph.D. diss., University of Wisconsin); see also Allen M. Hornblum, *Acres of Skin: Human Experiments at Holmesburg Prison* (New York: Routledge, 1998), a history of prisoners and medical experimentation.

113. Letter to the editor, Ludwig Gross, "Leukemic Blood Transfusion," *New York Times* (8 June 1949): 28.

114. "No Harm Results in Leukemia Test," *New York Times* (4 June 1950): 76. In *Acres of Skin,* Hornblum termed the Slater-Boy case a rarity based on the national attention it captured; however, it was only one of a number of "human interest" stories written about human guinea pigs during this period. For example, syphilis, atabrine, influenza, and arthritis experiments conducted on other Sing Sing inmates appeared in the *American Mercury* article

in 1954. Don Wharton, "Prisoners Who Volunteer Blood, Flesh, and Their Lives," *American Mercury* 79 (1954): 51–55.

115. Susan Lederer and Michael Grodin, "Historical Overview." In Michael A. Grodin and Leonard H. Glantz, eds., *Children as Research Subjects: Science, Ethics, and Law* (New York: Oxford University Press, 1994), 6.

116. In *Subjected to Science,* Lederer described the enormous outcry over children as research subjects. Animal protectionists organized the first societies for the prevention of cruelty to children, arguing that unrestricted animal testing would lead to similar practices on humans. The antivivisectionists feared that children would become "guinea pigs" for physicians and scientists engaged in nontherapeutic experiments.

117. Ruth R. Faden, Susan E. Lederer, and Jonathan D. Moreno, "US Medical Researchers, the Nuremberg Doctors Trial, and the Nuremberg Code: A Review of Findings of the Advisory Committee on Human Radiation Experiments," *Journal of the American Medical Association* 20/276 (27 Nov. 1996): 1668. Members of the Advisory Committee on Human Radiation Experiments determined that few physicians kept abreast of the Nuremberg doctors trial or applied the code to their own practices.

118. J. Stafford, "Reprieves Not Cures," *Science News Letter* (19 April 1952): 246.

119. Sidney Farber, "The Treatment of Acute Leukemia," [editorial] *Journal of Chronic Diseases* (April 1956): 455.

120. H. W. Dargeon, "Leukemia in Childhood: Current Therapeutic Considerations," *New York State Journal of Medicine* 56 (1 July 1956): 2079.

121. "Drug Lengthens Lives," *Science News Letter* (31 May 1952): 349.

Chapter Five • "Who's Afraid of Death on the Leukemia Ward?"

1. Peter De Vries, *The Blood of the Lamb* (Boston: Little, Brown, 1961).

2. W. J. Smith, [Book review], *Commonweal* 76 (20 April 1962): 93.

3. [Book review], *Times Literary Supplement* (18 May 1962): 353.

4. Raised by parents who had emigrated from the Netherlands, author Peter De Vries grew up in a Dutch-American, Calvinist community in the middle of Chicago. After graduating from Calvin College in Grand Rapids, Michigan, in 1931, he began working odd jobs and publishing his poetry and short stories. In the late 1930s, he became an associate editor and then coeditor of *Poetry.* He then moved to the *New Yorker* at the invitation of James Thurber, who had read his work and been interviewed by him at a lecture sponsored by *Poetry.* In 1943, he married Katinka Loeser, and the couple had four children between 1945 and 1952. Emily was born on 26 Oct. 1949. The family lived in Westport, Connecticut, and five novels published in the 1950s and early 1960s mocked the town's ways of life. This autobiographical sketch is based on the description of J. H. Bowden, *Peter De Vries* (Boston: Twayne, 1983), one of the few biographical sources on De Vries and his work.

5. Bowden, *Peter De Vries,* 75. Emily was the only child in the family honored by a dust jacket photograph.

6. Ibid., 9.

7. Ibid., 75.

8. Harold Dargeon, "Pediatrics at Memorial Hospital for Cancer and Allied Diseases," Record Group 160.7, Series 2, Box 1, Memorial Sloan Kettering Cancer Center Archives, Rockefeller Archive Center, New York.

9. Bowden, *Peter De Vries*, 75. In later years, he was invited to give academic lecture series, and he was elected to the National Institute of Arts and Letters in 1969 and the American Academy of Arts and Letters in 1983. (J. H. Bowden, 8). Edwin T. Bowden, the author of the definitive bibliography of De Vries's work, compared his best novels to those of other famed twentieth-century writers such as Evelyn Waugh and James Thurber. See Edwin T. Bowden, *Peter De Vries: A Bibliography, 1934–1977* (Austin: Humanities Research Center, University of Texas at Austin, 1978), 9.

10. Bowden, *Peter De Vries*, 20.

11. Tom and Alice Fleming, "Special Report: Cancer in Children," *Cosmopolitan* (Aug. 1963): 52–57; James C. G. Conniff, "The Brightening Outlook in Child Cancer," *Family Circle* (April 1962); "Children and Cancer," *Good Housekeeping* (February 1964): 40. The articles did not specify whether the American Cancer Society or another health organization had edited the pieces, but this seems likely based on the striking similarities between the articles.

12. Natcher Stewart, "Cancer in Children," *Health* 11/2 (Oct. 1965): 4.

13. Ibid.

14. Algernon B. Reese, "Heredity and Retinoblastoma," *Archives of Ophthalmology* 42 (Aug. 1949): 119–122. Reese studied 150 cases of sporadic retinoblastoma (both parents were healthy) and their siblings in order to determine its pattern of incidence. He determined that siblings only had a 4 percent chance of suffering from the disease, but seven of eight children of parents who had survived the disease had retinoblastoma.

15. Walter S. Ross, "What Parents Should Know about Childhood Cancer," *Reader's Digest* (March 1967): 6. According to a conference presentation by leading retinoblastoma researchers, the estimated cure rate for children in Group I, the earliest stage of disease, was 90 percent. Cured children were expected to have their vision completely restored and be free of cataracts. Survival decreased as the disease progressed and cure in Group V, the final stage, decreased to 20 percent. Algernon B. Reese and Robert M. Ellsworth, "Management of Retinoblastoma," *CA: A Cancer Journal for Clinicians* 14/1 (Jan.–Feb. 1964): 9, provided an abstract of their contribution to the 1964 ACS Scientific Session.

16. See Amy Swerdlow, *Women Strike for Peace: Traditional Motherhood and Radical Politics in the 1960s* (Chicago: University of Chicago Press, 1993), and Milton S. Katz, *Ban the Bomb: A History of SANE, the Committee for a Sane Nuclear Policy, 1957–1985* (New York: Greenwood Press, 1986), for detailed descriptions of these events and their imprint on American families. For a broader review of the atomic age, radiation, and health, see Susan M. Lindee, *Suffering Made Real: American Science and the Sufferers at Hiroshima* (Chicago: University of Chicago Press, 1994), and Paul Boyer, *By the Bomb's Early Light: American Thought and Culture at the Dawn of the Atomic Age* (New York: Pantheon Books, 1985).

17. Tom and Alice Fleming, "Special Report: Cancer in Children," *Cosmopolitan* (Aug. 1963): 53.

18. Ibid.

19. Ibid.

20. Ibid.

21. Charles Cameron, *The Truth about Cancer* (Englewood Cliffs, N.J.: Prentice-Hall, 1956), 232.

22. Ibid.

23. Ibid.

24. De Vries, *The Blood of the Lamb*, 169.

25. Ibid.

26. Ibid., 170.

27. Ibid., 174.

28. See Ilana Löwy, *Between Bench and Bedside: Science, Healing, and Interleukin-2 in a Cancer Ward* (Cambridge, Mass.: Harvard University Press, 1996), 57, for a concise review of the evolution of cooperative clinical groups in pediatric oncology. Two institutional histories are invaluable sources for this narrative: Kenneth M. Endicott, "The Chemotherapy Program," *Journal of the National Cancer Institute* 19 (1957): 283, and C. Gordon Zubrod, "Historic Milestones in Curative Chemotherapy," *Seminars in Oncology* 6 (1979): 490–505.

29. Stephen P. Strickland, *Politics, Science, and Dread Disease: A Short History of United States Medical Research Policy* (Cambridge, Mass.: Harvard University Press, 1972), 200.

30. Löwy, 58.

31. Joseph H. Burchenal, "Recent Advances and Perspectives in the Chemotherapy of Acute Leukemia." In *Proceedings of the 5th National Cancer Conference* (Philadelphia: J. B. Lippincott Co., 1965), 651–657; an updated and revised version of the article was published as Joseph H. Burchenal, "Treatment of the Leukemias," *Seminars in Hematology* 3/2 (April 1966): 122–131.

32. Charlotte Tan, Hideko Tasaka, Kou-Ping Yu, M. Lois Murphy, and David A. Karnofsky, "Daunomycin: An Antitumor Antibiotic in Treatment of Neoplastic Disease: Clinical Evaluation with Special Reference to Childhood Leukemia," *Cancer* 120/3 (1967): 333–353.

33. A. Goldin, J. M. Venditti, S. R. Humphreys, and N. Mantel, "Modification of Treatment Schedules in the Management of Advanced Mouse Leukemia with Amethopterin," *Journal of the National Cancer Institute* 17 (1956): 203–212.

34. Emil J. Freireich, Edmund Gehan, Emil Frie III, Leslie Schroeder, Irving J. Wolman, Rachad Anbari, E. Omar Burgert, Stephen D. Mills, Donald Pinkel, Oleg S. Selawry, John H. Moon, B. R. Gendel, Charles L. Spurr, Robert Storrs, Farid Haurani, Barth Hoogstraten, and Stanley Lee, "The Effect of 6-Mercaptopurine on the Duration of Steroid-Induced Remissions in Acute Leukemia: A Model for Evaluation of Other Potentially Useful Therapy," *Blood* 21 (1963): 699–716.

35. See Emil Frei III, "Intramural Therapeutic Research at the National Center Institute, Department of Medicine: 1955–1965," *Cancer Treatment Reports* 68/1 (Jan. 1984): 21–30.

36. See Emil Frei III and Emil J. Freireich, "Progress and Perspectives in the Chemotherapy of Acute Leukemia," *Advances in Chemotherapy* 2 (1965): 269–289, and O. Selawry, "New Treatment Schedule with Improved Survival in Childhood Leukemia," *Journal of the American Medical Association* 194 (1965): 75–81.

37. Vincristine was an alkaloid derived from the Madagascar periwinkle. It attacked rapidly proliferating cells like cancer cells, intestinal epithelium, and bone marrow. Thus, it was an effective agent with life-threatening side effects.

38. Frei and Freireich, "Progress and Perspectives in the Chemotherapy of Acute Leukemia";

Emil J. Friereich, Myron Karon, and Emil Frei III, "Quadruple Combination Therapy (VAMP) for Acute Lymphocytic Leukemia of Childhood," *Proceedings of the American Association for Cancer Research* 5 (1964): 20; and James F. Holland, "Formal Discussion: The Clinical Pharmacology of Anti-Leukemia Agents," *Cancer Research* 25 (1965): 1639–1641. Freireich recalled that the other research groups participating in the cooperative trials organized by the National Cancer Center were already involved with other protocols and unwilling to transfer their patients to the VAMP study, but he believed it had so much promise that he withdrew from the cooperative group and enrolled all newly admitted National Cancer Center patients into VAMP.

39. The VAMP protocol cured three patients. In December 1985, the cover of *Cancer Research* featured a family photograph alongside a graph displaying ALL survival in children younger than twenty treated by Cancer and Leukemia Group B. J.C.G., the initials of the woman in the family portrait, had been treated with VAMP at the Clinical Center of the National Cancer Center when she was seven years old. Twenty-three years later, she remained cancer-free and had three young children of her own.

40. Oncovin and Purinethol are the trademarked names of these drugs and were used in the regimen acronyms.

41. De Vries, *The Blood of the Lamb*, 175.

42. Ibid., 177.

43. Ibid., 179.

44. Ibid., 193.

45. See James A. Whiteside, Fred S. Philips, Harold W. Dargeon, and Joseph H. Burchenal, "Intrathecal Aminopterin in the Neurological Manifestations of Leukemia," *Archives of Internal Medicine* 101 (1958): 280, and Emil J. Frei, "The Effectiveness of Combinations of Anti-Leukemic Agents in Inducing and Maintaining Remission in Children with Acute Leukemia," *Blood* 26 (1965): 642–656. It was later demonstrated that, if intrathecal methotrexate injections were given earlier, they significantly reduced the risk of meningeal leukemia. In the 1970s, low-dose irradiation coupled with methotrexate injections reduced its incidence to less than 10 percent.

46. In 1960, a syndicated medical advice columnist offered information on finding treatment for leukemia. The physician wrote that many mothers of leukemic children asked him whether they should scour the country for the latest therapeutic innovation. The physician recommended readers from small cities to travel to a larger city with a university hospital, Mayo Clinic, or Memorial Hospital to get advice and advised those residing in large cities to find a hematologist. Walter C. Alvarez, "Advice Offered on Searches for Treatment of Leukemia," *Houston Post* (27 June 1960).

47. Abraham B. Bergman and Charles J. A. Schulte, eds., "Care of the Child with Cancer," *Pediatrics* 40 (Sept. 1967) (Supplement): 487–546.

48. John R. Hartmann, Panel Discussion, "Care of the Child with Cancer," *Pediatrics* 40 (Sept. 1967) (Supplement): 546.

49. Charles Q. McClelland, Panel Discussion, "Care of the Child with Cancer," *Pediatrics* 40 (Sept. 1967) (Supplement): 543.

50. Mila Pierce, Panel Discussion, "Care of the Child with Cancer," *Pediatrics* 40 (Sept. 1967) (Supplement): 545

51. Charles J. A. Schulte III, "Programs of the Cancer Control Program: U.S. Public Health Service," *Pediatrics* 40 (Sept. 1967) (Supplement): 527.

52. Charles Q. McClelland, "Relationship of the Physician in Practice to a Children's Cancer Clinic," *Pediatrics* 40 (Sept. 1967) (Supplement): 537.

53. Michael B. Rothenberg, "Reactions of Those Who Treat Children with Cancer," *Pediatrics* 40 (Sept. 1967) (Supplement): 510.

54. John R. Hartmann, "The Physician and the Children's Cancer Center," *Pediatrics* 40 (Sept. 1967) (Supplement): 523.

55. For a history of the professionalization of hematology and oncology and the contested relationship between the two specialties, see Keith Wailoo, *Drawing Blood: Technology and Disease Identity in Twentieth-Century America* (Baltimore: Johns Hopkins University Press, 1997); and Gretchen Krueger, "'Where Does Hematology End and Oncology Begin?': Questions of Professional Boundaries and Medical Authority," *Journal of Clinical Oncology,* 24/16 (1 June 2006): 2583–2588, and "The Formation of the American Society of Clinical Oncology and the Development of a Medical Specialty, 1964–1973," *Perspectives in Biology and Medicine* (fall 2004): 537–551.

56. Denman Hammond, "Panel Discussion," *Pediatrics* 40 (Sept. 1967) (Supplement): 541.

57. Farber also urged parents to use large research centers rather than local doctors for the coordination of their child's cancer care. He instructed patients to visit a research center for an initial consultation in order to make a plan of treatment to be continued at home under the local physician's supervision. He noted, "It is doubtful whether more than ten thousand (of 260,000 cancer patients who will die during the year) of these will be systematically treated and studied by doctors skilled in cancer chemotherapy" (56). Another specialist in childhood cancer (unnamed in the article) agreed with Farber's advice, saying, "I do not mean to slight the skill and compassion of the local doctor, but he cannot possibly maintain a general practice and keep up with the latest developments in such a rapidly growing field as chemotherapy." Tom and Alice Fleming, "Special Report: Cancer in Children," *Cosmopolitan* (Aug. 1963): 58.

58. John R. Hartmann, "The Physician and the Children's Cancer Center," *Pediatrics* 40 (Sept. 1967) (Supplement): 526.

59. Ibid.

60. Angela B. Tonyan, "Role of the Nurse in a Children's Cancer Clinic," *Pediatrics* 40 (Sept. 1967) (Supplement): 532.

61. John Laszlo, *The Cure of Childhood Leukemia: Into the Age of Miracles* (New Brunswick, N.J.: Rutgers University Press, 1995), 143.

62. Ibid.

63. Hartmann, "The Physician and the Children's Cancer Center," 524.

64. Ibid.

65. De Vries, *The Blood of the Lamb,* 170.

66. Ibid., 173.

67. After Carol's death, at the end of the novel, Wanderhope found an audiocassette on which Carol revealed, "I might as well say that I know what's doing on." Despite her father's efforts, she had learned of the identity and prognosis of her illness. Ibid., 241.

68. Ibid., 202.

69. For a glimpse of both sides of the debate in 1962, see Victor A. Gilbertsen and Owen H. Wangensteen, "Should the Doctor Tell the Patient That the Disease Is Cancer?" *CA: Cancer Journal for Clinicians* 12/3 (May–June 1962): 82–86, and Lemuel Bowden, "Editor's Interview: The

Patient with Incurable Cancer," *CA: Cancer Journal for Clinicians* 12/3 (May–June 1962): 104–106.

70. D. Oken, "What to Tell Cancer Patients: A Study of Medical Attitudes," *Journal of the American Medical Association* 175 (1961): 1120–1128, and Howard Witzkin and John D. Stoeckle, "The Communication of Information about Illness," *Advances in Psychosomatic Medicine* 8 (1972): 185–189, illustrated this dramatic shift.

71. Albert J. Solnit and Morris Green, "Psychological Considerations in the Management of Deaths on Pediatric Hospital Services. 1. The Doctor and the Child's Family," *Pediatrics* 24 (1959): 106–112. In 1962, George T. Pack ("Counseling the Cancer Patient: Surgeon's Counsel," *CA: Cancer Journal for Clinicians* 12/6 [Nov.–Dec. 1962]: 211–212) wrote that he preferred not to discuss the nature of surgery or consequent disabilities with his child patients. This was in stark disagreement with children's books designed to prepare children for doctor's appointments and hospitalization that were published beginning in the 1950s. Pack recommended that surgeons comfort the young patient after the recovery from anesthetic and encourage him to be "brave" and "courageous" for his parents.

72. Morris Green, "Care of the Dying Child," *Pediatrics* 40 (Sept. 1967) (Supplement): 495.

73. See M. Lois Murphy, "Acute Leukemia." In Sydney S. Gellis and Benjamin M. Kagan, eds., *Current Pediatric Therapy* (Philadelphia: W. B. Saunders, 1964), 275–279.

74. Joel Vernick and Myron Karon, "Who's Afraid of Death on a Leukemia Ward?" *American Journal of the Diseases of Children* 109 (May 1965): 393.

75. Ibid.

76. Ibid., 395.

77. Ibid., 396.

78. Editorial, "What Should the Child With Leukemia Be Told?" *American Journal of the Diseases of Children* 110 (Sept. 1965): 231.

79. Vernick and Karon, "Who's Afraid of Death on a Leukemia Ward?" 394.

80. Editorial: "What Should the Child With Leukemia Be Told?" *American Journal of the Diseases of Children* 110 (Sept. 1965): 231.

81. Response: Joel Vernick and Myron Karon, "What Should the Child with Leukemia Be Told?" *American Journal of the Diseases of Children* 110 (Sept. 1965): 335.

82. Response: Henry F. Lee, "What Should the Child With Leukemia Be Told?" *American Journal of the Diseases of Children* 110 (Dec. 1965): 704.

83. Ibid.

84. Response: Alfred Hamady, "What Should the Child with Leukemia Be Told?" *American Journal of the Diseases of Children* 110 (Dec. 1965): 704.

85. Martin Pernick explored these classic arguments in the history of truth telling in "Childhood Death and Medical Ethics: A Historical Perspective on Truth Telling in Pediatrics," *Progress in Clinical Biological Research* 139 (1983): 173–188.

86. Doris A. Howell, "A Child Dies," *Journal of Pediatric Surgery* 1/1 (February 1966): 2–7 (first issue), and reprinted in *Seminars in Hematology* 3/2 (April 1966): 168–173. Howell discussed the role of the physician specifically during each phase of acute leukemia from telling diagnosis until after death.

87. Paul Chodoff, Stanford B. Friedman, and David A Hamburg, "Stress, Defenses and Cop-

ing Behavior: Observations in Parents of Children with Malignant Disease," *American Journal of Psychiatry* 120 (February 1964): 743–749.

88. Stanford B. Friedman, "Care of the Family of the Child with Cancer," *Pediatrics* 40 (Sept. 1967) (Supplement): 499.

89. Ibid.

90. De Vries, *The Blood of the Lamb*, 226.

91. Ibid., 222.

92. Ibid., 219.

93. Ibid., 184.

94. Ibid., 183.

95. Ibid., 206.

96. Ibid., 225.

97. "An American Mother Moves Next to Sainthood," *Life* 54 (29 March 1963): 38–39. Seton's story was covered widely in the press through articles in *U.S. News and World Report, Newsweek, Time,* and the *Saturday Evening Post.* Before Mother Seton could be designated a saint, two more miracles needed to be performed in her name and authenticated by the Vatican Congregation of Rites.

98. Ibid., 38.

99. Chester M. Southam, Lloyd F. Craver, Harold W. Dargeon, and Joseph H. Burchenal, "A Study of the National History of Acute Leukemia with Special Reference to the Duration of the Disease and the Occurrence of Remissions," *Cancer* 4 (1951): 39–59.

100. Joseph H. Burchenal and M. Lois Murphy, "Long-Term Survivors in Acute Leukemia," *Cancer Research* 25 (1965): 1491–1494. He presented the data and addressed criticisms of the data at a symposium on the clinical aspects of acute leukemia and Burkitt's tumor in Boston, Massachusetts, on 20 September 1967. The talk was published as Joseph H. Burchenal, "Long-Term Survivors in Acute Leukemia and Burkitt's Lymphoma," *Cancer* 21/4 (April 1968): 595–599. By carefully tracking the patients in this cohort, Burchenal also tentatively made suggestions regarding when it was appropriate to discontinue treatment, an issue debated at the time.

101. Ibid.

Chapter Six • "The Truly Cured Child"

1. Amy Louise Timmons, "Is It So Awful?" *American Journal of Nursing* 75/6 (June 1975): 988, and Amy Louise Timmons, "Is It So Awful?" *Journal of Pediatrics* 88/1 (Jan. 1976): 147–148.

2. Remissions were now indicated by less than 5 percent abnormal cells in the peripheral blood and bone marrow and the restoration of normal marrow function as demonstrated by adequate numbers of white blood cells, red blood cells, and platelets.

3. See R. J. Amin Aur, J. V. Simone, H. O. Husto, M. S. Verzosa, and D. Pinkel "Cessation of Therapy during Complete Remission of Childhood Acute Lymphocytic Leukemia," *New England Journal of Medicine* 291 (5 Dec. 1974): 1230–1234, for an assessment of the proper length of treatment to prevent relapse. Aur and his colleagues at St. Jude Children's Research Hospital regarded aggressive chemotherapy, central nervous system irradiation, and two to three years of

complete remission as the optimum treatment schedule. Other research groups continued therapy for as long as five years.

4. Researchers at St. Jude pioneered the addition of craniospinal irradiation to combination chemotherapy regimens in their "total therapy" studies. Donald Pinkel, "Five-Year Follow-Up of 'Total Therapy' of Childhood Lymphocytic Leukemia," *Journal of the American Medical Association* 216 (1971): 648–652, and Joseph Simone, "Total Therapy Studies of Acute Lymphocytic Leukemia in Children: Current Results and Prospects for Cure," *Cancer* 30 (1972): 1488–1494.

5. For a review of the microbial agents that cause disease in patients with hematologic malignancy and the prevention or treatment of the complications, see Arthur S. Levine, Robert G. Graw, and Robert C. Young, "Management of Infections in Patients with Leukemia and Lymphoma: Current Concepts and Experimental Approaches," *Seminars in Hematology* 9/2 (April 1972): 141–179.

6. Emil J. Freireich, "The Best Medical Care for the 'Hopeless' Patient," *Medical Opinion* 8/2 (February 1972): 55.

7. "Cancer Remains Major Killer of School Children," *Journal of the American Medical Association* 234/2 (13 Oct. 1975): 140.

8. Ibid., 140.

9. J. L. Young, H. W. Heisis, E. Siberverg, and M. H. Myers, "Cancer Incidence, Survival and Mortality for Children under 15 Years of Age," archives, American Cancer Society, Atlanta, Georgia, 1976.

10. Ibid. Other five-year rates were as follows: 40 percent for brain tumors, 27 percent for both medulloblastoma and neuroblastoma, 63 percent for lymphoma, 59 percent for kidney tumors, 24 percent for osteogenic sarcoma, 17 percent for Ewing's sarcoma, 89 percent for eye cancer, and 39 percent for connective tissue cancers.

11. Joseph Simone, "Acute Lymphocytic Leukemia in Childhood," *Seminars in Hematology* 11 (Jan. 1974): 26–27.

12. "Children Winning More Cancer Battles, But War Isn't Won," *News American* (11 Sept. 1978), Archives, American Cancer Society.

13. Emil J. Freireich, "The Best Medical Care for the 'Hopeless' Patient," *Medical Opinion* 8/2 (February 1972), 54.

14. Ibid.

15. Ibid., 55.

16. George J. Annas, *Rights of Hospital Patients: The Basic ACLU Guide to Hospital Patient's Rights* (New York: Avon Press, 1975), and George J. Annas, Leonard H. Glantz, and Barbara F. Katz, *Informed Consent to Human Experimentation* (Cambridge, Mass.: Ballinger, 1974). Also see Renèe C. Fox, *Experiment Perilous: Physicians and Patients Facing the Unknown* (Glencoe, Ill.: Free Press, 1959), for a landmark sociological study of a metabolic investigation. Fox observed "the talented, young, academically-inclined physicians who undertook the experiment and the articulate, relatively young patients, ill with serious, chronic, and often unusual conditions who underwent it" (24).

17. Ilana Löwy, *Between Bench and Bedside: Science, Healing and Interleukin-2 in a Cancer Ward* (Cambridge, Mass.: Harvard University Press, 1996), 280. I have modified the factors that she applied to oncology generally to encompass the close relationship between pediatric cancer care and clinical experimentation.

18. See Ruth R. Faden and Tom L. Beauchamp, in collaboration with Nancy M. P. King, *History and Theory of Informed Consent* (New York: Oxford University Press, 1986), for a discussion of medical experiments involving children and the requirements of informed consent for minors.

19. Paul Ramsey, *The Patient as Person: Explorations in Medical Ethics* (New Haven, Conn.: Yale University Press, 1970), 11–12. Informed consent required that patients had adequate information, comprehend the information, and make a voluntary decision about their care or participation in a clinical trial. Ramsey referred to the use of children in research by proxy consent as a "prismatic case" to clearly define the meaning of the consent requirement (35).

20. For examples of improper research conduct with children, see M. H. Pappworth, *Human Guinea Pigs* (Boston: Beacon Press, 1968), and William J. Curran and Henry K. Beecher, "Experimentation in Children," *Journal of the American Medical Association* 210 (6 Oct. 1969): 77–83.

21. Melvin J. Krant, Joseph L. Cohen, and Charles Rosenbaum, "Moral Dilemmas in Clinical Cancer Experimentation," *Medical and Pediatric Oncology* 3 (1977): 146.

22. Ibid., 142–143.

23. Harold Y. Vanderpool, "The Ethics of Experimentation with Anticancer Drugs." In Steven C. Gross and Solomon Garb, eds., *Cancer Treatment and Research in Humanistic Perspective* (New York: Springer, 1985), 16–46. See also T. L. Beauchamp and J. F. Childress, *Principles of Biomedical Ethics* (New York: Oxford University Press, 1979).

24. Ida Marie Martinson, *Home Care for the Dying Child: Professional and Family Perspectives* (New York: Appleton-Century-Crofts, 1976), 25.

25. The Cancer Centers Program, Our History, *www.cancer.gov.*

26. Robin Frames, "Cancer Centers Help Families with Children," *Logan, Utah Herald Journal* (2 July 1975). Robin Frames, "At Cancer Centers for Kids, Total Care Dispensed," *Florida Times-Union* (10 July 1975), Archives, American Cancer Society.

27. Susan Spence Moe, "For Children with Cancer, There's Hope," *News and Observer* [Raleigh, North Carolina] (4 April 1976), Archives, American Cancer Society.

28. Dave Anderson, "Kim's Houses," *New York Times* (11 Nov. 1979), 3.

29. James O. Clifford, "Hospital Involves Family in Cancer Fight," Archives, American Cancer Society.

30. K. Forte, "Pediatric Oncology Nursing: Providing Care through Decades of Change," *Journal of Pediatric Oncology Nursing* 18/4 (July–Aug. 2001): 154–163; S. P. Hiney and F. M. Wiley, "Historical Beginnings of a Professional Nursing Organization Dedicated to the Care of Children with Cancer and Their Families: The Association of Pediatric Oncology Nursing from 1974–1993," *Journal of Pediatric Oncology Nursing* 13/4 (Oct. 1996): 196–203; Patricia Greene, "Acute Leukemia in Children," *American Journal of Nursing* 75/10 (Oct. 1975): 1711.

31. Victoria Graham, "The Sad Wait at Ricky's House," *San Jose Mercury News* (5 Jan. 1975), Archives, American Cancer Society.

32. Ibid.

33. "Leukemia Strains Emotional Ties," *Medical World News* (6 April 1973): 23. The study was conducted from 1967 to 1972.

34. Ibid. As of 2008, there are twenty-four National Cancer Center–designated cancer centers and thirty-nine National Cancer Center–designated comprehensive cancer centers across the country.

35. Wilbur acknowledged that the "traditional" treatment was "probably characteristic of most centers providing care for severe childhood illness today." Thus, Kaplan's results did apply to a significant number of patients and families receiving treatment in the early 1970s.

36. Graham, "The Sad Wait at Ricky's House."

37. John J. Spinetta, David Rigler, and Myron Karon, "Anxiety in the Dying Child," *Pediatrics* 52/6 (Dec. 1973): 841–844.

38. Myra Bluebond-Langner, *The Private Worlds of Dying Children* (Princeton, N.J.: Princeton University Press, 1978). In the book, Bluebond-Langner constructed a five-act play titled "The World of Jeffrey Andrews" to present a summary of her observations in a format that highlighted children, the primary actors in her research.

39. In his inflammatory work, Illich called for a revolutionary change in health care delivery. Ivan Illich, *Medical Nemesis: The Expropriation of Health* (London: Calder and Boyars, 1975).

40. Boston Women's Health Book Collective, *Our Bodies, Our Selves: A Course By and For Women* (Boston Women's Health Book Collective and the New England Free Press, 1971). To embed *Our Bodies, Our Selves* in its broader historical context of activism and women's health, see Sandra Morgen, *Into Our Own Hands: The Women's Health Movement in the United States, 1969–1990* (New Brunswick, N.J.: Rutgers University Press, 2002).

41. For an account of earlier challenges regarding the use of alternative medicines in cancer therapy, see Barbara N. Clow, *Negotiating Disease: Power and Cancer Care, 1900–1950* (Montreal: McGill-Queen's University Press, 2001). As in the case of Johnny Gunther's diet therapy, some parents of young cancer sufferers welcomed alternative therapies because they provided a less expensive, milder option, they could be administered at home, or they served as a last resort when standard chemotherapy or medical procedures had failed. In the 1970s, laetrile, a derivative of apricot pits, was a popular therapy. John A. Richardson and Patricia Griffin, *Laetrile Case Histories: The Richard Cancer Clinic Experience* (Westlake Village, Calif.: American Media, 1977), is a tract written by the therapy's leading proponents. The laetrile debate intensified when the parents of Chad Green, a young acute leukemia sufferer, demanded to administer laetrile as their son's primary treatment, not as a complementary therapy or as a palliative agent during the terminal stage of cancer. See Marion Steinman, "A Child's Fight for Life: Parents vs. Doctors," *New York Times Magazine* (Dec. 10, 1978): 160, for an extended discussion of the legal fight. For contemporary studies of alternative medicine and pediatric cancer patients, see Cathy Faw, Ron Ballentine, Lois Ballentine, and Jan van Eys, "Unproved Cancer Remedies: A Survey of Use in Pediatric Outpatients," *Journal of the American Medical Association* 238/14 (3 Oct. 1977): 1536–1538; and Thomas W. Pendergrass and Scott Davis, "Knowledge and Use of 'Alternative' Cancer Therapies in Children," *American Journal of Pediatric Hematology/Oncology* 3/4 (winter 1981): 339–345.

42. Elisabeth Kübler-Ross, *On Death and Dying: What the Dying Have to Teach Doctors, Nurses, Clergy, and Their Own Families* (New York: MacMillan, 1969), preface.

43. Ibid., 11.

44. Elisabeth Kübler-Ross, *On Children and Death* (New York: Macmillan, 1983).

45. L. Wainwright, "Profound Lesson for the Living," *Life* (21 Nov. 1969): 36–43. Also see Barney G. Glaser and Anselm L. Strauss, *Awareness of Dying* (Chicago: Aldine, 1965); David Sudnow, *Passing On: The Social Organization of Dying* (Englewood Cliffs, N.J.: Prentice-Hall, 1967); Philippe Ariès, *Western Attitudes toward Death: From the Middle Ages to the Present* (Baltimore:

Johns Hopkins University Press, 1974); and Renée Fox, *Essays in Medical Sociology* (New York: John Wiley and Sons, 1979), for further contemporary discussion of the denial of death and the construction of medical environments that isolated the dying and dehumanized the dying process. Refer to Peter G. Filene, *In the Arms of Others: A Cultural History of the Right-to-Die in America* (Chicago: Ivan R. Dee, 1998), for a recent analysis of changing practices around death and dying.

46. Kenneth L. Woodward, "Living with Dying," *Newsweek* 91/18 (1 May 1978): 52–63. Robert Kastenbaum labeled the surge in attention to death and dying as the death-awareness movement. See Jessica Mitford, *The American Way of Death* (New York: Simon and Schuster, 1963).

47. "A Better Way of Dying," *Time* 111/23 (5 June 1978): 66.

48. J. Fishhoff and N. O'Brien, "After the Child Dies," *Journal of Pediatrics* 88/1 (Jan. 1976): 140–146.

49. Cicely Saunders and Mary Baines, *Living with Dying: The Management of Terminal Disease* (New York: Oxford University Press, 1983); Sandol Stoddard, *The Hospice Movement: A Better Way of Caring for the Dying* (New York: Stein and Day, 1978); and a more modern evaluation and criticism of hospice and its changing goals, Cathy Siebold, *The Hospice Movement: Easing Death's Pains* (New York: Twayne, 1992).

50. Ida Marie Martinson, *Home Care for the Dying Child: Professional and Family Perspectives* (New York: Appleton-Century-Crofts, 1976), 5.

51. Ida M. Martinson, "Why Don't We Let Them Die at Home? *RN* (Jan. 1976): 58.

52. Martinson, *Home Care for the Dying Child*, 9.

53. Emily Kulenkamp and Ida M. Martinson, *Eric* (Minneapolis: University of Minnesota School of Nursing, 1974), 2. Martinson's interactions with two young patients and their families were published in two slim pamphlets. The second was David N. Wetzel and Ida M. Martinson, *Meri* (Minneapolis: University of Minnesota School of Nursing, 1975).

54. *Journal of Allied Health* (1976), 26

55. Wetzel and Martinson, *Meri*, 29.

56. Martinson, *Home Care for the Dying Child*, 37.

57. Barbara Etzel, "The Role of Advocacy in the Rite of Passage." In Ida Marie Martinson, *Home Care for the Dying Child*, 55.

58. Ida M. Martinson, "The Child with Leukemia: Parents Help Each Other," *American Journal of Nursing* 76/7 (July 1976): 1121. Participants shared their experiences in pamphlet form as well as in a parental support groups. The article described another group at Babies Hospital, Children's Medical and Surgical Center, Columbia-Presbyterian Medical Center, New York.

59. I. M. Martinson, D. Geis, M. A. Anglim, E. Peterson, M. Nesbit, J. Kersey, "When the Patient Is Dying: Home Care for the Child," *American Journal of Nursing* 77/11 (1977): 1815–1817.

60. I. M. Martinson, M. Palta, and N. Rude, "Death and Dying: Selected Attitudes and Experiences of Minnesota's Registered Nurses," *Nursing Research* 9 (April 1977): 197–206.

61. Kathy Forte, "Pediatric Oncology Nursing: Providing Care through Decades of Change," *Journal of Pediatric Oncology Nursing* 18/4 (July–Aug.): 154–163.

62. D. Pinkel, "Treatment of Acute Leukemia," *Pediatric Clinics of North America* (February 1976): 128.

63. I. H. Krakoff, "The Case for Active Treatment in Patients with Advanced Cancer: Not Everyone Needs a Hospice" *CA: Cancer Journal for Clinicians* 29/2 (1979): 108–111.

64. Robert W. Buckingham, *The Complete Hospice Guide* (New York: Harper, 1983), 85.

65. In 2002, the Institute of Medicine recommended a new model of care in its groundbreaking report *When Children Die: Improving Palliative and End-of-Life Care for Children and Their Families* (Washington, D.C: National Academies Press, 2002). This integrated system ensured that children had access to a hospice team and eliminated the need for a six-month prognosis. Under this plan, hospices could be reimbursed when providing palliative care to children who were continuing to receive treatment intended to cure their disease or prolong their lives. This plan may succeed because it combines key elements from Martinson's early efforts yet also acknowledges the unique meanings of children and the aggressive orientation that characterizes pediatric oncology in the United States.

66. Stephen Hess, "Letter of Transmittal," White House Conference on Children, 1970, *Report to the President, White House Conference on Children* (Washington, D.C., 1971), 10.

67. Beatrice Gross and Ronald Gross, *The Children's Rights Movement: Overcoming the Oppression of Young People* (Garden City, New York: Anchor Press/Doubleday, 1977). The authors also promoted organizations such as the Children's Defense Fund and documents such as the United Nations Declaration of the Rights of the Child as methods for expanding the rights of a child. The Grosses were regarded as leaders in educational change in the 1970s, and they combined their academic interests with a devotion to social change by publishing widely both in their fields and for the public.

68. Robert and Suzanne Massie, *Journey* (New York: Alfred A. Knopf, 1975). The Massies published this description of their son Bobby's struggle with hemophilia and the family's battle to provide care once their son turned eighteen. At the time of his diagnosis, the prognosis was grim—54 percent of sufferers died before the age of five and only 11 percent lived to age twenty-one. Refer to Jacquie Gordon, *Give Me One Wish* (New York: W. W. Norton, 1988), for cystic fibrosis. The average life expectancy of a child with cystic fibrosis was only five years when Christine was born in 1961 and had risen to nineteen years when she died at age twenty-one in 1983. Like the Massies, Gordon included critiques as well as praise for aspects of her child's care.

69. Jonathan B. Tucker, *Ellie: A Child's Fight against Leukemia* (New York: Holt, Rinehart, and Winston, 1982). The volume was actually a composite of three different case histories. The author also used sources from Candlelighters, the National Cancer Center, the American Cancer Society, the Leukemia Society of America, the Children's Hospital of Philadelphia, and many individual physicians to construct the single narrative. Eric Lax, *Life and Death on 10 West* (New York: Times Books, 1984), was based on the young patients treated in the new bone marrow transplantation unit at the UCLA Medical Center.

70. Ilana Löwy, *Between Bench and Bedside: Science, Healing and Interleukin-2 in a Cancer Ward* (Cambridge, Mass.: Harvard University Press, 1996), 73–83.

71. Carol Kruckenberg, *What Was Good about Today* (Seattle: Madrona, 1984). Kruckenberg was a mother of an eight-year-old suffering from acute myelobrastic leukemia, a cancer that killed 95 percent of its sufferers within a year. The reader gains a complete knowledge about the features of a children's hospital and its response to the needs of parents. In Ray Erbol Fox, *Angela Ambrosia* (New York: Alfred A. Knopf, 1979), Angela's father discusses how their personal preferences were considered and valued during her treatment at Sloan-Kettering.

72. Barron H. Lerner, *The Breast Cancer Wars: Hope, Fear, and the Pursuit of a Cure in Twentieth-Century America* (New York: Oxford University Press, 2001), esp. 141–169.

73. Joan E. Fretwell, "A Child Dies," *Nursing Times* 69/27 (5 July 1973): 867–871, is an exception. Fretwell criticizes the uncaring attitudes displayed by physicians and other associated health personnel involved in her thirteen-year-old daughter's case. Marcia Friedman, *The Story of Josh* (New York: Praeger, 1974), focuses on the oral record of the sufferer himself but also addresses the high cost of medical care and the problems of hospital bureaucracy.

74. Doris Lund, *Eric* (Philadelphia: J. B. Lippincott, 1974). A comprehensive list of illness narratives specific to childhood cancers can be found in Hazel B. Benson, ed., *The Dying Child: An Annotated Bibliography* (Westport, Conn.: Greenwood Press, 1988). Benson separated narratives published between 1960 and 1987 into those written by mothers, fathers, and siblings.

75. Lund, *Eric,* 15.

76. Ibid.

77. Ibid., 58.

78. Ibid., 106.

79. Elaine Ipswitch, *Scott Was Here* (New York: Delacorte, 1979). The story of fifteen-year-old Scott's experiences with Hodgkin's disease was also excerpted in Elaine Ipswitch, "Maybe My Time Is Up," *Reader's Digest* 115/689 (Sept. 1979): 147–151. The book and article describe Scott's and the family's daily struggles to cope with the disease and their discovery (in journal entries read after his death) that he was aware of his condition and impending death. Also see Mary Winfrey Trautmann, "The Absence of the Dead," *Parents Magazine* 60/2 (February 1985): 83, 160, 162, 164, 166–68; Terry Pringle, *This Is the Child* (New York: Alfred A. Knopf, 1983); Rose Levit, *Ellen: A Short Life Long Remembered* (San Francisco: Chronicle Books, 1974); Nancy Roach, *The Last Day of April* (New York: American Cancer Society, 1974); and Owenita Sanderlin, *Johnny* (Cranbury, N.J.: Barnes, 1968), for examples of this style and content.

80. Refer to Mari Brady, *Please Remember Me: A Young Woman's Story of Her Friendship with an Unforgettable Fifteen-Year-Old Boy* (Garden City, N.Y.: Doubleday, 1977), for another narrative that highlighted the concerns of adolescents. The collection of illness narratives published during this period includes infants through eighteen-year-olds. Many narratives about older children include the words of the young patient as captured in diary entries, letters, or recordings.

81. Lund, *Eric,* 32.

82. Rachel Carson, *Silent Spring* (Boston: Houghton Mifflin, 1962). In *Before Silent Spring: Pesticides and Public Health in pre-DDT America* (Princeton, N.J.: Princeton University Press, 1974), James Whorton describes criticisms of hazardous chemical contaminants prior to publication of Carson's classic work and examines why they did not incite the same level of fervor.

83. The childhood cancer clusters were also reported in popular literature such as Natcher Stewart, "Cancer in Children" *Health* 11/2 (published by the American Osteopathic Association) (Oct. 1965): 4–10.

84. A viral agent was suspected in the Niles case because of contemporary research on a new type of lymphoma identified by researcher Denis Burkitt, an Irish government surgeon working in Uganda. After Burkitt discovered that the cancer cases spanned Africa, he sought an explanation for this narrow band of disease. Striking similarities between yellow fever maps and the new tumor map suggested that an insect vector such as the anopheles mosquito was responsible for transmitting a lymphoma virus. Cooperative work between Burkitt and investigators at Memorial Hospital held promising implications for members of the leukemia and can-

cer research community for two reasons; it suggested that a direct link existed between viruses and human cancers and demonstrated that therapy with low doses of existing chemotherapeutic agents could stimulate tumor regression and even cure advanced tumors. For an overview of this joint work, see Joseph H. Burchenal and Denis P. Burkitt, eds. *Treatment of Burkitt's Tumor: Proceedings of a Conference Organized by the Chemotherapy Panel of the International Union against Cancer* (Berlin: Springer-Verlag, 1966), and J. H. Burchenal, "Geographic Chemotherapy: Burkitt's Tumor as a Stalking Horse for Leukemia: Presidential Address" *Cancer Research* 26/12 (Dec. 1966): 2393–2405.

85. Paula DiPerna, *Cluster Mystery: Epidemic and the Children of Woburn, Mass.* (St. Louis: C. V. Mosby, 1985), 30.

86. In 1978, Michael Brown of the *Niagara Gazette* wrote an investigative report on Hooker Chemical Company dumping into the Love Canal area near Buffalo, New York. Gibbs continues to coordinate a campaign against environmental contaminants called the Citizens Clearinghouse for Hazardous Waste.

87. Paula DiPerna, *Cluster Mystery: Epidemic and the Children of Woburn, Massachusetts* (St. Louis: C. V. Mosby, 1985). Phil Brown and Edwin J. Mikkelsen, *No Safe Place: Toxic Waste, Leukemia and Community Action* (Berkeley: University of California Press, 1990), focuses on the important role of community members in moving the case into a national forum.

88. DiPerna, *Cluster Mystery*, 66.

89. Ibid., 114.

90. The legal disputes in the Woburn case were described in Jonathan Harr's best-selling book *A Civil Action* (New York: Random House, 1995).

91. DiPerna, *Cluster Mystery*, 182.

92. James Kloss, "Brain Tumor: Disease without Early Warning," *Chicago Daily News* (13 Jan. 1975).

93. See two institutional histories for a chronology of St. Jude's establishment and activities: *A Dream Come True: The Story of St. Jude Children's Research Hospital and ALSAC* (Dallas: Taylor, 1983) and *From His Promise: The Story of St. Jude Children's Research Hospital and American Lebanese Syrian Associated Charities* (Memphis: Guild Bindery Press, 1996).

94. "Maintaining a Normal Life." In *Proceedings of the First National Conference for Parents of Children with Cancer, 23–25 June 1978*. U.S. Department of Health and Human Services, National Cancer Institute, NIH Publication No. 80-2176 (June 1980): 1.

95. Ibid.

96. Jordan R. Wilbur, "Parent as Part of the Treatment Process." In *Proceedings of the First National Conference for Parents of Children with Cancer*, 155–162.

97. Sister Margaret Weeke (moderator), "Coping: Teenage Panel Discussion," In *Proceedings of the First National Conference for Parents of Children with Cancer*, 51–88.

98. J. V. Simone, R. J. A. Aur, H. O. Hustu, M. S. Verzosa, and D. Pinkel, "Three to Ten Years after Cessation of Therapy in Children with Leukemia," *Cancer* 42 (1978): 839.

99. G. J. D'Angio, "Complications of Treatment Encountered in Lymphoma-Leukemia Long-Term Survivors," *Cancer* 42 (1978): 1022.

Conclusion

1. "Girl Fights Bone Cancer. Father Hoped to Save Her From Being Crippled," *Brooklyn Eagle* (27 Sept. 1939), Memorial Hospital 1939–1940 Scrapbook, SKCC, RAC.

2. Ibid.

3. No title, Jamaica (New York) Long Island Press (20 Sept. 1939), Memorial Hospital 1939–1940 Scrapbook, SKCC, RAC. The story was also distributed by the Associated Press and appeared in newspapers from Baltimore to West Virginia and Pennsylvania. Dorothy's mother died not long before the girl's diagnosis, so the decision to allow the operation was left solely to her father.

4. "Valiant Is the Word," *New York Journal American* (11 Oct. 1939): 6–7, Memorial Hospital 1939–1940 Scrapbook, SKCC, RAC.

5. Edward Kennedy case was described in "Teddy's Ordeal," *Time* (3 Dec. 1973): 86.

6. Unlike osteogenic sarcoma (cancer of the bone) that often rapidly spread to the lung and had only a five-year survival rate of 5–23 percent when an immediate amputation had successfully removed the primary tumor, chondrosarcoma had a 70 percent survival rate. It is unclear why Kennedy was included in the study. Perhaps his exact diagnosis was uncertain at the time.

7. N. Jaffe, E. Frei III, D. Traggis, and Y. Bishop, "Adjuvant Methotrexate and Citrovorum-Factor Treatment of Osteogenic Sarcoma," *New England Journal of Medicine* 291 (1974): 994–997. In the second study, the Acute Leukemia Group B had tested the efficacy of adriamycin in osteosarcoma in patients without detectable metastases. Preliminary data had demonstrated that adriamycin was active in a wide spectrum of neoplastic diseases including patients with pulmonary metastases from osteosarcoma. Researchers decided to conduct a larger study in which they would administer the drug during a period when the patient's body burden of tumor cells was the lowest. Thus, children were given the drug two weeks after the surgical amputation of the primary lesion. The drug delayed gross metastases. See Engracio P. Cortes, James F. Holland, Jaw J. Wang, Lucius F. Sinks, Johannes Blom, Hanjurg Senn, Arthur Bank, and Oliver Glidewell, "Amputation and Adriamycin in Primary Osteosarcoma," *New England Journal of Medicine* 291 (1974): 998–1000.

8. In the late 1970s, a study published by Gerald Rosen of Memorial Sloan-Kettering in 1978 found that administering drug therapy before surgery caused bone tumors to shrink and killed cancer cells that could have spread throughout the body. Rosen reported that the preoperative chemotherapy increased four-year survival time from 50 to 70 percent. See "Treating Bone Cancer in Children," *San Francisco Chronicle* (12 Sept. 1978) for a summary.

9. "Valiant Is the Word," *N.Y. Journal American* (11 Oct. 1939): 6, Memorial Hospital 1939–1940 Scrapbook, SKCC papers, RAC.

10. National Cancer Center–supported Cancer Clinical Trials, www.cancer.gov/clinical trials/facts-and-figures/page2.

11. Robert and Suzanne Massie, *Journey* (New York: Alfred A. Knopf, 1975), xi.

12. Ibid., 76.

13. See John E. O'Malley, "Psychiatric Sequelae of Surviving Childhood Cancer," *American Journal of Orthopsychiatry* 49/4 (Oct. 1979): 608–616; Judith W. Ross, "Social Work Intervention with Families of Children with Cancer: The Changing Critical Phases," *Social Work in Health*

Care 3/3 (spring 1978): 257–272; Lynn Kagen-Goodheart, "Reentry: Living with Childhood Cancer," *American Journal of Orthopsychiatry* 47/4 (Oct. 1977): 651–658.

14. Stories of negotiation between physicians, parents, and the courts over the refusal of treatment or the use of alternative treatments for childhood cancers are regularly in the news. Such contests have flourished with the growth of consumer-oriented medicine, the greater involvement of parents in daily cancer care, and the persistence of aggressive, multiphase therapy. For a recent case, see Bernadine Healy, "The Tyranny of Experts," *U.S News and World Report* (27 June 2005). The blog http://prayforkatie.blogspot.com/ includes links to court documents, fundraising events, and media coverage from the case as well as regular postings from Katie Wernecke, the thirteen-year-old girl with lymphoma who is at the center of the debate.

15. See www.alexslemonade.com/aboutthestand.php.

16. "Girl Who Sold Lemonade for Cancer Research Dies at 8," *Seattle Times* (online version), 3 Aug. 2004.

17. For history and critical commentary on telethons, see Paul Longmore, "The Cultural Framing of Disease: Telethons as Case Study," *PMLA* 120/2 (March 2005): 502–508. Longmore points to a telethon aired in New York in 1949 by the Damon Runyon Cancer Fund as the first program produced by a private, voluntary health charity designed to reach the growing audience now accessible through broadcast television.

18. Lisa Belkin, "Charity Begins at . . . The Marketing Meeting, The Gala Event, The Product Tie-In: How Breast Cancer Became This Year's Cause," *New York Times Magazine* (22 Dec. 1996): 42.

INDEX